European Defense Research & Development
New Visions & Prospects for Cooperative Engagement

Jeffrey P. Bialos & Stuart L. Koehl, Editors

March 2004

© Center for Transatlantic Relations, Johns Hopkins University 2004

Center for Transatlantic Relations
The Paul H. Nitze School of Advanced International Studies
The Johns Hopkins University
1717 Massachusetts Avenue N.W., Suite 525
Washington, DC 20036
Tel. (202) 663-5880
Fax (202) 663-5879
Email: transatlantic@jhu.edu
http://transatlantic.sais-jhu.edu

Table of Contents

List of Acronyms and Abbreviations v

Editors' Preface and Acknowledgements ix

Chapter 1
   European Defense R&D in Perspective:
      The Nexus to Defense Transformation,
      Coalition Warfare and European Security Policy 1
   Jeffrey Bialos & Stuart Koehl

Chapter 2
   European National Directives on R&D: United Kingdom
      R&D Policy and Transatlantic Cooperation 37
   David Gould

Chapter 3
   European National Directions on R&D—
      New Processes and Approaches 47
   Yves Boyer

Chapter 4
   Leveraging European Dual Use Technologies
      for Defense Needs 59
   Kenneth Flamm

Chapter 5
   The EU's Role in Leveraging Multiple-Use
      Technologies for Defense Needs 79
   Klaus Becher

Chapter 6
   European Defense Research & Technology (R&T)
      Cooperation: A Work in Progress 95
   Andrew James

Chapter 7
   International Defense R&D Cooperation:
   From Competition to True Cooperation—
   The Case of US-Japan Defense R&D
   Cooperation in Transition 117
   Masako Igekami

Appendix A
   Symposium Agenda and List of Participants 141

Appendix B
   Sweden's Approach to Defense Research
      and Transformation 147
   Staffan Nasstrom

Appendix C
   Dual Use Technology in European Space
      Research and Development 163
   Daniel Hernandez

About the Authors 173

List of Acronyms and Abbreviations

ACT	Allied Command Transformation
ATC	Allison Technical Company
BDM	Ballistic Missile Defense
BOA	*Bulle Operationnelle Aero Terrestre*
C4ISR	Command, Control, Communications, Computers, Intelligence, Surveillance and Reconnaissance
CERN	European Council for National Research
CEPA	Common European Priority Areas
CFSP	Common Foreign and Security Policy
COST	Cooperation in the Field of Scientific Technical Research
COTS	Commercial Off-the-Shelf
CPCO	*Centre de Preparation et de Conduite des Operations*
DARPA	Defense Advanced Research Project Agency
DDR&E	Director of Defense Research and Engineering
DERA	Defense Engineering and Research Agency
DGA	*Delegation Generale pour L'Armement*
DoD	Department of Defense
DPA	Defense Procurement Agency
DTSI	Defense Trade Security Initiative
EADS	European Aeronautical, Defense and Space Company
ECAP	European Capacity Action Plan
ENIF	Experimental Network Integration Facility
EREA	European Research Establishments in Aeronautics
ERG	European Research Grouping
ERRF	European Rapid Reaction Force
ESA	European Space Agency
ESDP	European Security and Defense Policy
ETAP	European Technology Acquisition Program
EU	European Union
EUCLID	European Cooperation for Longterm in Defense
FDS	Future Rapid Effects System
FOAS	Future Offensive Air System
FOFA	Follow on Forces Attack
FMV	Swedish Defense Acquisition Agency
GARTEUR	Group for Aeronautical Research and Technology in Europe

GMES	Global Monitoring for the Environment and Security
GOP	Group of Personalities
ICOC	International Code of Conduct
ICOG	International Cooperative Opportunities Group
IEPG	Independent European Programme Group
IFSEC	Industry Forum on Security Cooperation
IHI	Ishikawajima Hanma Heavy Industries
IP	Internet Protocol
IPSC	Protection and Security of the Citizen
ITAP	Integrated Technology Acquisition Plan
JDA	Japanese Defense Agency
JFC	Joint Forces Control
JSF	Joint Strike Fighter
JUEP	Joint UAV Experimentation Program
KUR	Key User Requirements
LOI	Letter of Intent
LPM	*Loi de Programmation Militaire*
MEADS	Medium Range Air Defense System
MIC	Multinational Interoperability Council
MIDS	Multifunctional Information Distribution System
MoD	Ministry of Defense
MTCR	Missile Technology Control Regime
MTDP	Midterm Defense Program
MOU	Memorandum of Understanding
NBC	Nuclear Biological and Chemical
NBD	Network-Based Defense Initiative
NCW	Network Centric Warfare
NEC	Network Enabled Capability
NITEworks	Network Integration Test & Experimentation
NRF	NATO Response Force
OA	Operations Analysis
OCCAR	*Organization Conjoint de Cooperation en Matiere d'Armament*
OMG	Operational Maneuver Group
PFI	Privately Financed Initiative
PJHQ	Permanent Join Headquarters
R&D	Research and Development
R&T	Research and Technology
R&TD	Research and Technology Development
RA&D	Requirement Analysis and Defense

RDT&E	Research, Development, Testing & Evaluation
RMA	Revolution in Military Affairs
S&TF	System and Technology Forum
SACEUR	Supreme Allied Commander Europe
SDI	Strategy Defense Initiative
STOVL	Short Take Off/Vertical Landing
TA	Technical Agreements
TDCs	Transatlantic Defense Companies
THALES	Technology Arrangements for Laboratories for Defense European Science
UAV	Unmanned Air Vehicles
UCAV	Unmanned Air Combat Vehicle
UOR	Urgent Operational Requirements
VAATE	Versatile Advanced Affordable Turbine Engines
VHSIC	Very High Speed Integration Circuit
VLSI	Very Large Scale Integrated Circuit
WEAG	Western European Armaments Group
WEAO	Western European Armaments Organization
WEU	Western European Union

Editors' Preface & Acknowledgements

European defense research and development is a dynamic subject, with significant changes in institutions, processes, and focus in recent months as the security environment itself is being recast. In Europe today, the question of what research and development is needed to facilitate the acquisition of military capabilities relevant to address 21st century threats is very much on the agenda. The role of the European Union in this field is rapidly being shaped, questions of national policy are being considered, and there is an increased focus on the need to match requirements, resources, and supporting research efforts and to proceed more rapidly, efficiently and, ultimately, effectively with respect to the insertion of new technologies and capabilities into Europe's fielded forces.

Therefore, we are pleased to provide this volume, which addresses in depth the issues surrounding European defense research and development and provides a broad range of European, American and Asian perspectives on these matters.

On behalf of the Center for Transatlantic Relations at the Paul H. Nitze School of Advanced International Studies, we would like to thank those people and institutions that have contributed to the preparation of this volume. First, thanks go to the speakers and other participants at the Symposium on "European Defense Research and Development: New Visions and Prospects for Cooperative Engagement" ("Symposium"), held at the Center on June 6, 2003; the participants included a range of senior European and U.S. government, military, academic and business representatives. A number of the Symposium speakers subsequently contributed essays or presentations to this volume. Also, many of the ideas and discussions at the Symposium have been reflected in this volume. Appendix A sets forth the Symposium agenda and the list of Symposium speakers. The power point presentations made at the Symposium are available on the website of the Center on Transatlantic Relations (http://transatlantic.sais-jhu.edu).

Second, we acknowledge the yeoman like efforts of Dara Iserson, Christian Liles and Simona Marin for their assistance in the preparation and editing of this volume. Finally, we extend our thanks to Dr. Daniel Hamilton, Ms. Katrien Maes, Ms. Jeannette Murphy and the staff of the Center for Transatlantic Relations for their support and assistance in the preparation for and support during the Symposium.

While the insights and assistance of all of these people have helped to make the volume what it is, please note that the responsibility for its contents are exclusively the province of the editors and authors and not the institutions with which they are affiliated.

Jeffrey P. Bialos
Stuart L. Koehl
Washington, D.C.
Spring 2004

Chapter 1

European Defense R&D In Perspective: The Nexus to Defense Transformation, Coalition Warfare and European Security Policy

Jeffrey Bialos
Stuart Koehl

This volume provides a range of critical perspectives on key issues with respect to European defense research and development (R&D). Specifically, it evaluates:

- the current state and future directions of largely separate European national research efforts as European nations grapple with new and transformational visions of warfare;
- the European capability to leverage its cutting edge and well funded civilian research for defense needs;
- the prospects for European coalescence in R&D as the European Union builds on pre-existing cooperative research efforts, seeks to forge a new European Armaments and Research Agency and close the well known capabilities gap with the United States; and
- the realities of and prospects for Transatlantic engagement in research and development in support of coalition war fighting as Europe and the United States struggle to find common ground after recent policy divergences such as Operation Iraqi Freedom.

This volume contains a series of works dealing with each of these topics. It reflects the existence of consensus on some but not all of the issues concerning European Defense research and development. While the emergence of a European defense identity, including a research and development function, now appears widely accepted in principle in Europe, European views are mixed on other key issues:

- Around what future visions should European R&D be organized? Should Europe develop and follow a vision of military transformation similar to that now in vogue in the United States, or does this risk too much dependence on U.S. technology and command, control, communications and intelligence architecture?

- Should European priorities focus on capabilities for high intensity missions such as those envisaged for the new NATO Response Force or for low intensity Petersberg tasks (peacekeeping and the like) identified by European Headline Goals? Will NATO or the European Union be the center of gravity for European capabilities acquisition and related R&D activities?

- Is Transatlantic R&D cooperation desirable? And if so, what would be the appropriate forms for Transatlantic R&D cooperation and broader security engagement?

A Frame of Reference for Thinking About European Defense R&D

These divergent views highlight that European defense R&D cannot be viewed in a vacuum; defense R&D is a means to an end, not an end in itself. Thus, European defense R&D is derivative of overall concepts of European security and warfare, the missions and capabilities around which European forces are being designed, and the overarching nature of European cooperation and Transatlantic engagement in defense.

When viewed in this broader geopolitical and war fighting context, the question for the future is whether and to what degree European coalition partners can obtain affordable, technically proficient, innovative and interoperable war fighting capabilities needed to meet 21st century security threats.

To achieve this "armaments" goal and have an effective R&D program, Europe must take many steps. Thus, as many participants at the recent R&D Symposium agreed, Europe, *acting as Europe*, should:

- Develop organizing principles—that is, European military requirements which reflect an evaluation of the actual and projected capabilities of potential adversaries and what capabilities are needed to defense against these threats;

- Increase its defense spending, including spending for defense R&D, and shift its spending toward these new requirements and away from old priorities;

- Take steps to foster a framework which facilitates greater competition in European defense markets, which has proven to be a key to providing both innovation and affordability; and

- Determine the extent to which interoperability with the United States is an important goal, and the degree to which R&D should be allocated to this task.

Europe faces the challenge of addressing these issues in a dynamic security context:

Changing Threats: In an era of U.S. conventional military predominance and the absence of any peer competitor, potential adversaries have sought to develop asymmetrical means of leveraging their limited resources and exploiting our perceived weaknesses. Thus, we have witnessed the emergence of a range of new threats, including terrorism as a preferred method of warfare, missiles of all types—cruise missiles, ballistic missiles, and shoulder fired projectiles, and systematic national efforts to develop weapons of mass destruction.[1] In the future, unmanned air vehicles (UAVs)—a potentially available technology in an era of globalization and technological diffusion—also may prove a potent threat. This range of capabilities can be viewed as the preferred means of attack by rogue nations and sub-national groups—as in effect, "poor man's weapons" against the overwhelming military power of the United States and its allies.

U.S. Transformation: Changing threats have resulted in considerable changes in U.S. defense strategy and force planning. What is now described as defense "transformation"—a term that has many meanings—and what was previously described as the revolutions in "business" and "military" affairs reflects the U.S. Department of Defense's

---

[1] For an overview of the missile threat, see Jeffrey P. Bialos and Stuart H. Koehl, "Forging a Transatlantic Consensus on Missile Defense", *International Spectator*, Vol.XXXVIII No.4, Oct.-Dec. 2004.

approach to adapting its war fighting and business operations to the 21st century. At its heart, transformation includes a thrust toward greater reliance on technology and a force structure that is more network-centric in nature, with greater lethality and precision, and the ability to operate seamlessly both among U.S. services and with potential coalition partners.

On the U.S. side, these concepts are gradually being translated into force reality—And today are better integrated into defense operations than armaments R&D and procurement. For a variety of reasons, acquisition patterns are slower to change in response to changes in operational method. Slowly, however, the nature of defense demand is changing in the aftermath of the Iraq war, with the acceleration of key transformational programs, the cancellation of legacy programs, and the movement of funding away from modernization of existing programs. These changes are also accompanied by changes in our processes for doing defense R&D—where the concept of "spiral development," among others, are being utilized in an effort to facilitate more rapid movement from the drawing board to the war fighter.

Defense Industrial Consolidation: While government laboratories may undertake a significant portion of basic defense research, the development and application of technology to military systems is largely the realm of private industry. In recent years, the Transatlantic defense industry has undergone significant change, including large-scale consolidation in response to a decade of defense budget downsizing. Europe is also moving towards the privatization of government owned and controlled defense companies; gradually governments are recognizing that putting defense firms on a commercial footing and subjecting them to market forces yields efficiencies, more innovation and better leveraging of commercial technology in an era where breakthrough technologies often flow from the commercial to the defense realm. At the same time, however, industrial consolidation does raise serious questions about maintaining competition and the innovation it fosters. Transformation requires innovation, and innovation is in part a product of competition. How that plays out is a key issue for the future of defense R&D.

Changing War Fighting Coalitions. During the Cold War, the United States expected to fight its wars within the context of formalized alliances and organized coalitions such as NATO. With the collapse of the Soviet Union, the common perception of the threat which provided cohesion to the alliance has diminished. The emergence of multi-polar regional threats has forced both the United States and European countries to work within a framework of shifting, ad hoc coalitions—what Secretary of Defense Rumsfeld has called "coalitions of the willing," partners whose capabilities and methods may vary widely from those of other members of the traditional coalition. In recent military engagements, the United States has tended to work with such coalitions rather than through formal existing coalition structures such as NATO. At the same time, alternative centers of leadership and organization in the security arena are emerging—most notably the European Union (EU), which is forging its own European Security and Defense Policy (ESDP). The effect that these developments will have on Transatlantic security and defense cooperation, including cooperative R&D, cannot yet be fully comprehended.

The Future of NATO: The Iraq War revealed fundamental differences in outlook and preferred strategy between the United States and some of the major NATO allies, creating stress fractures within the alliance that threaten the future viability of NATO as a collective security alliance. Already some defense analysts are speaking of the "myth" of NATO, or NATO as a discussion group rather than a real military alliance. While the invocation of Article 5 after the terrorist attacks of 11 September 2001 demonstrated NATO's solidarity in the face of post-Cold War era threats, the limited participation of NATO in the subsequent war and differences over Operation Iraqi Freedom have led to real questions concerning NATO's future as a military alliance—especially for out of area operations. The shared political will for such activities is not high and the degree of force inter-operability remains low.

* * *

In sum, these and other dynamic developments will have considerable impacts on European defense R&D. This reflects the fact that there is a complex interrelationship between geopolitics, coalition warfare, economics, and armaments development and cooperation. As a general proposition, over time, nations that do not cooperate in armaments development and procurement are less likely to fight together and remain allies—and *vice versa*.

European R&D Investment: The Nexus to the "Capabilities Gap"

Within this broader security context, defense R&D clearly has and does play a critical role in the development of military capabilities on both sides of the Atlantic.

When viewed in historical context, the Transatlantic military capabilities gap is in large measure a function of two fundamental realities: 1) the importance of technology in modern warfare; and 2) the enormous difference between European and US defense R&D, both in absolute levels and the manner in which Europe invested over the last decade—i.e., either on a fragmented and duplicative national basis or a cooperative basis that focused on *"jus retour"* and national work share rather than "best value."

The Role of Technology in Modern Warfare

If nothing else, recent military history highlights that technology matters significantly on the battlefield (at least during the high-intensity phases of conflict). The wars in Afghanistan and Iraq (and to a lesser extent in Bosnia and Kosovo) have shown how rapidly technological developments are affecting the transformation of warfare. Witness, for example, the use of UAVs not only for reconnaissance, but for targeting, long-term surveillance, communications relay, and even as lethal platforms; the use of the Global Navigation System for navigation, fire control and weapons guidance; and the real-time integration of sensors and shooters.

R&D Investment Levels: Past, Present & Future

Significantly, U.S. leadership in almost all of these areas can be traced to its sizable and sustained investment in defense R&D in the last two decades and its willingness, from both a strategic and technological standpoint, to leverage the so-called information age-driven Revolution in Military Affairs in transforming its force capabilities. Simply put, the European investment in defense R&D has lagged significantly behind the United States in both scope and magnitude—producing serious gaps in core military capabilities.

Figure 1 below highlights the enormous historic disparity, which continues to exist today and in the near future, between U.S. and European defense R&D spending. The United States has spent somewhere from 3 to 5 times as much on its defense investment than Europe over time.

When viewed in per capita terms, the R&D spending disparity is stark. As **Figure 2** shows, the United States spends almost three times as much per solider/sailor/airman on R&D as do the leading European and Asian powers. Smaller European countries lag even further behind.

Looking forward, a continuation of the U.S.-European defense investment disparity is likely and will, in all likelihood, perpetuate and exacerbate the Transatlantic capabilities gap unless European governments are willing to take strategic actions to change the current trajectory.

Significantly, despite considerable public attention to the investment and capabilities gap, the United States continues to spend much more on defense R&D than Europe, in both absolute and relative terms. Indeed, the trend lines in Figure 1 are revealing. Since September 11, the U.S. spending on defense R&D has increased significantly while

Figure 1. U.S. vs European Defense R&D Budgets, 1992-2004

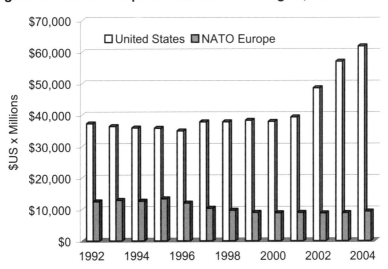

Source: IISS Military Balance

Figure 2. R&D Spending per Capita

Source: National Science Foundation

Europe's spending has essentially remained flat. Moreover, in the future, U.S. R& D spending is likely to continue to increase modestly or at least remain relatively stable while Europe's ability to increase its investment will likely be modest due to spending constraints imposed by European realities (including the limitations of the Maastricht Treaty) and the need to utilize most funds for the modernization of legacy forces as well as operations and maintenance demands.

Figure 3 highlights the dilemma faced by European governments. It tracks spending on military modernization, defined as improvements to existing systems and development of new systems/capabilities. Significantly, the United States is matched only by the UK in terms of investment per soldier. It shows that the United States and the United Kingdom are the only NATO countries seriously investing in force modernization. Other countries are investing much less both absolutely and proportionally. This modernization spending "wedge" reflects exploding personnel and O&M costs in most European countries. Indeed, the vicious circle of spending funds to maintain oversized forces and aging legacy systems is one faced both in the United States and in Europe. However, the United States has sufficient fund to both maintain its legacy forces while investing for the future. In Europe, in a constrained spending environment, funding requirements for legacy forces significantly curtail investment in future systems.

In short, on its current trajectory, Europe appears locked into this vicious cycle that leaves few resources for investment. Moreover, the fragmented nature of the spending and focus on "national" work share in cooperative programs that do exist exacerbates the situation and creates tendencies toward "second best" development efforts with limited results. If Europe is to escape the difficult dilemma in which it finds itself, it must select from a series of difficult strategic actions that require real societal trade-offs. Europe either must: increase its defense top line budgets to free up funds for greater R&D; specialize its forces and R&D efforts (similar to what Israel had done in focusing on key war fighting areas for its investments); or reprioritize its defense spending and reorganize its defense R&D budgets to spend what it has more effectively (through force downsizing and a shift toward new war fighting concepts). The reality is that more defense spending is considered unlikely in the coming years due to competing domestic needs, the Maastricht Treaty limits, and other spending constraints. Europe's challenge, therefore, is to spend better. The only way to meaningfully do this is to cut existing forces and redirect the

Figure 3. Defense Modernization Funding, US. vs. Europe

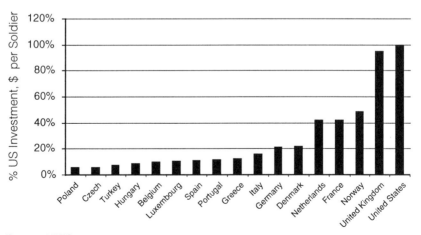

Source: NATO

funds released towards R&D investment and defense transformation. Taking these steps will not be easy, and will involve hard choices such as cutting force structures, legacy programs, and unneeded infrastructure. As discussed below, a number of European nations are embarked on promising defense reform efforts; while the pace and scope has been slow it does seem to be gaining speed and traction.

Ultimately, the nature and degree of European R&D over the next generation will determine whether and to what extent Europe can develop robust military capabilities that can effectively address 21st century security threats and close the existing capabilities gap. To do this, however, Europe must not only get its broader security strategy in order, but must work to organize a more efficient and effective research and development program as one element of this overall strategy.

European "National" R&D: The Emergence of Transformation & Expeditionary Warfare As Organizing Principles

As European governments seek to free up funding for future investment in defense R&D, a threshold question is "invest in what?" Given the significant resource limitations, European governments cannot afford to invest under a "let 1,000 flowers bloom" approach, and must make difficult choices that focus on projects and programs directly related to producing future capabilities. At the heart of the issue are whether and to what degree European nations are embracing concepts of defense transformation, including a movement toward network-centric warfare, as part of their vision for future war fighting capabilities and, therefore, as organizing principles for R&D.

While there are differences between European and U.S. conceptions of military transformation, the key nations in Europe (reflected in the size and scope of their defense R&D efforts) are nevertheless moving forward—to different degrees and at different paces—toward fundamental changes in war fighting concepts. On a fundamental level, there is now a recognition in Europe of the need to shift toward more mobile, rapid tempo expeditionary forces for out of area conflicts (rather than the traditional reliance on geographically-fixed forces) for either high intensity or Petersburg missions. In a certain sense, given the Europe's historic Cold War focus on large, heavy armored-mechanized ground forces as the backbone of its defense, this fundamental change in philosophy is more central in Europe than the United States. As a corollary, European governments are also moving toward the use of network-centric capabilities and improvements in technology, including in sensors and long-range precision strike. The recent string of technologically-driven American military successes in Iraq, Afghanistan, and in the Balkans (at least during the high intensity phase of conflict) has had a significant effect on European military thinking and created more "demand" for these types of capabilities within European military establishments. These new directions are gradually being reflected in national R&D programs as governments seek to translate these new approaches into practical solutions for the war fighter.

Not surprisingly, the European reception of these new concepts has been mixed. The United Kingdom and Sweden are at the forefront followed by France, more recently Germany, and to a limited degree by Italy. While the UK and Swedish concepts are now in the process of implementation through R&D and fielding of war fighting capabilities, the actual efforts are more limited in France and Italy (focused on specific projects) and still are largely at the "power point engineering" phase in Germany.

The United Kingdom. In Chapter 2 of this volume, David Gould, Deputy Chief Executive of the United Kindom's Defense Procurement Agency outlines the UK's emerging approach to R&D in his essay on *European National Directives on R&D: United Kingdom R&D Policy and Transatlantic Cooperation.* Not surprisingly, the United Kingdom, is probably closest to the U.S. model of transformation, embracing not only a move toward network-centric warfare but toward new business models for R&D and procurement that favor competition, public/private partnerships, the increasing role of commercial off-the-shelf, and reliance on "dual use" technology where possible. Recognizing the importance of defense industries in its investment strategy, the UK Ministry of Defense (MoD) has developed an industrial policy that maintains a strong emphasis on competition while seeking a more appropriate risk/reward ratio for industry and maximizing investment in research and technology (T&T).[2] The

---

[2] See David Gould, "UK Defense Procurement: Fighting on Two Fronts" (Briefing to the Center for Transatlantic Relations), January 2003

UK also is a leader in privatization of defense industrial capabilities within Europe. Most recently, the UK privatized a significant portion of the Defense Engineering and Research Agency (DERA), its defense laboratory and research apparatus—in recognition of the fact that putting these capabilities on a commercial footing can encourage both innovation and efficiency.

At the heart of the United Kingdom's transformational efforts is its concept of "network enabled capability." As explained by the UK MoD, this means "[l]inking sensors, decision makers, and weapons systems to that information can be translated into synchronized and overwhelming military effect at optimum tempo."[3] While this sounds very similar to American concepts of "sensor to shooter" connectivity, there are on close examination key differences. In particular, recognizing its resource constraints, the UK focus is on developing a "plug and play" architecture that can be interoperable with, and plug into other networks (i.e., the United States, NATO or EU nations) rather than operate its own full spectrum of network-centric capabilities.

This approach is reflected in a range of UK national and cooperative R&D programs. This includes the Bowman Tactical Communications System (which is being designed with a view to being interoperable with the U.S. Joint Tactical Radio System, the next generation U.S. secure communications program), the Unmanned Air Combat Vehicle (UCAV) and other unmanned air vehicle programs, as well as a range of precision air-delivered munitions.

Also central to the UK approach is its philosophy that "military advantage is generally enjoyed by those nations with access to high technology" and that investment in R&D is critical to the prosperity of its defense industry. These concepts are reflected in the UK's Defence Industry Policy announced in October 2002, which focuses on sustaining technology in the supplier base, maintaining competition in defense markets, and facilitating partnerships with industry that allows MoD access to key industrial assets and intellectual property. Significantly, the UK policy considers foreign defense firms as part of the UK industrial base where they are engaged in design, development and production activities in the UK.

---

[3] United Kingdom Ministry of Defense, Network Enabled Capability: Working Definition, at www.mod.uk/issues/nec/working_definition.htm.

As Mr. Gould explains, other key UK thrusts include a focus on specific technologies (concentrating on areas where others are not) and investment in "niche" technologies and the need to more quickly pull technology into the supply chain. Thus, the UK MoD's new approaches include:

- integrated technology acquisition plans (ITAP) that focus on technology insertion at the earliest opportunities;

- a science and technology strategy that links up research and resources with planned capability outcomes and seeks measurable outputs; and

- rapid test and evaluation of ideas and prototypes—which one might call "test a little, deploy a little, test a little more."

Examples of this new approach include the Joint UAV Experimentation Program (JUEP), which involves both the user and customer community and focuses on the rapid operational use of assets. Specifically, JUEP calls for the development of a common concept of operations for UAVs, the development of new UAV payloads, exploration of UAV employment in maritime roles, and the resolution of issues related to UAV operations in controlled airspace, with the overall objective of informing new, emerging requirements.

Sweden. At the last year's Symposium, Major General Staffan Nasstrom, Chief Operational Manager of Defense Materiel Administration at the Swedish Ministry of Defense articulated the way in which Sweden is tackling transformation; his presentation is attached as Appendix B. As a small and sparsely populated country with limited resources and a distinct non-aligned defense policy, Sweden has shifted from a high degree of self-reliance toward more of a "participatory" approach that includes R&D cooperation (government-to-government and industrially) and results in mutual interdependence in relying on international capabilities and talents for its security of supply and innovation.

On balance, Sweden had adopted perhaps the most radical military reform program in Europe, significantly reducing force structure and legacy systems—on the order of some 75% — in order to invest in new technology and retaining only a vestigial capability in many areas deemed non-essential in the proposed transformed military. In

"Operation Leave Behind," surplus equipment (including reserve stockpiles and spare parts) was divested in a series of seven rounds of sales on the international market spanning the years 2001-2007. Redundant bases were closed. Net proceeds from sales and base closings were pumped back into the transformational process.[4] Defense industrial capacity supporting superfluous capabilities was transformed to other purposes, sold, or closed. With close cooperation between all the armed services, and centralized direction and coordination by the Swedish Defense Acquisition Agency (FMV), the Swedish Ministry of Defense hopes to complete four major demonstrations by 2007, in order to field the first generation, end-to-end network centric defense capability by 2010.

At the heart of Sweden's efforts is its "Network-Based Defense Initiative." In a new approach—which is similar to the U.S. Advanced Concepts Technology Demonstration program, Sweden has shifted toward the use of "demonstration-based acquisition" in its effort to acquire network centric capabilities. Specifically, Sweden is building a brass board system for each set of transformational capabilities (called "services"). The brass board systems are then tested in a series of advanced war fighting experiments, each of increasing complexity, until and end-to-end demonstration of the entire system has been accomplished. At each stage of the process, the Swedish military retains the residual war fighting capability in the brass board system, while the brass board serves as the baseline for operational system development. The first set of demonstrations occurred in 2002, with additional demonstrations schedules for 2003, 2005 and 2006, with initial operational capability deployment scheduled for 2010. Demonstration 2005 will focus on the critical capability of Battlespace Awareness, while Demonstration 2006 will address Command & Control Services.

Sweden's transformation focus can be seen in its resource allocation. Out of $180 million for NBD, approximately half is invested in C4I technology, with other large components in communications systems (19%), sensors (16%) and electronic warfare (19%). Interestingly, only 2% is for unmanned vehicles, which does not mean that Sweden is

---

[4] See "Phase-Out—Essential for Modernization," *Military Technology, Special Issue 2002*, Vol. XXVI, p.39.

ignoring this important transformational area, but rather that, having focused transformation on its own operational needs, the UAV systems it is developing, while technologically innovative, are relatively inexpensive platforms—new technology need not be costly. In addition, payload and communications systems for UAVs are covered under other technology areas (i.e., sensors, C4I, and communications), and thus the UAV investment does not reflect total system cost.

Sweden's model of R&D also reflects an openness to new technology, regardless of its origin, and a desire to participate more fully in parallel U.S. transformational programs. As a small country, Sweden recognizes that it cannot be industrially self-sufficient and has experienced significant defense industrial downsizing. While Sweden realizes that it has a competitive advantage in certain technology areas, and tends to direct its R&D funds there, in other areas it intends to rely on foreign technology and systems as a means to reduce cost and risk. To a large extent, the Swedish emphasis on Commercial Off-the-Shelf (COTS) technology both facilitates and reinforces the need for international partnerships. Thus, at the industrial level, Sweden has entered into partnerships with a number of British, Norwegian, Finnish and American companies, and high-level discussions have been held regarding technology transfer and information exchange between the U.S and Sweden, with the dual aims of technology transfer and information sharing.

Germany. In January 2004, German Defense Minister Peter Struck announced a far-reaching plan to reorganize the *Bundeswehr* around a series of new missions, including: conflict prevention; crisis management (including the fight against terrorism); support of alliance partners; homeland security rescue and evacuation of German citizens abroad; and practical assistance.[5]

More specifically, the post-reform *Bundeswehr* would be reorganized into:

- A rapid deployment force of 35,000 soldiers, that would be used for NATO high intensity missions (including service in the new NATO Response Force), EU peacekeeping operations, roles;

---

[5] See "Armed forces reform for the 21st century starting to take shape," Press release, German Embassy, London, United Kingdom (January 14, 2004), at www.german-embassy.org.uk.

- A stabilization force of 70,000 soldiers used for peacekeeping missions such as those currently being carried out in Afghanistan and the Balkans; and

- A support force of 137,500 soldiers for support and logistics for all missions, basic operations, and training.

The change contemplated is fundamental in nature and involves a shift away from 40 years of a heavy army designed to protect against and repel a Red Army invasion through Central Europe toward a smaller, multi-dimensional force that includes a modern expeditionary element—the rapid deployment force—for out of area actions from peacekeeping to high intensity missions. In effect, the German "transformation" efforts will be focused around this rapid expeditionary force, which could be used for operations like Iraqi Freedom or peacekeeping missions. It will include a focus on mobility, reconnaissance, and net-centric operations and a strong priority on inter operability in command, control, and communications systems with both EU and NATO forces. It also will recognize a new primacy on air power rather than the traditional German focus on ground warfare.

The plan also involves cuts in German troop levels (from 285, 000 to 250,000), the closing of 100 military bases and termination or reductions in a number of weapons acquisition programs (such as the Multi-Role Armored Vehicle being developed jointly by Rheinmetal of Germany and GKN of the UK). The plan would cut German defense spending overall by $32 billion pounds and would make the armed forces more professional.

While the German MoD plans to review and reorganize its R&D and procurement around its new missions, there are at present no details of what changes may be in store. In effect, the plans are in "powerpoint" form, but are only beginning to be implemented. At this writing, for example, it is uncertain how much funding will be freed up for new transformational purposes. Minister Struck has also indicated that the cuts would open up new funding for investment—but beginning in 2012, over seven years from now.[6]

---

[6] See Walker, Martin, "Walker's World: The NEW German Army," United Press International (January 14, 2004), at www.upi.com.

Certainly, one centerpiece of German "transformation" will be a prominent place for UAVs. German armaments officials have what some have called an infatuation with using UAVs for various intelligence, reconnaissance and surveillance missions and plan to replace various existing capabilities with UAV based capabilities. The recent EUROHAWK demonstration (which combines a U.S. Global Hawk with various sensor payloads developed by EADS) is part of this effort.

The new German plan, coming after earlier modest reform proposals, inevitably includes some compromises and seeks a balance between transformation toward new forces and continued maintenance of some legacy forces. Its troop reductions are modest, some programs like Eurofighter and MEADs will not be cut, and the German military draft will apparently be retained at least for now. Moreover, the reaction to it is mixed. Some believe it does not go far enough. The *Berliner Zietung*, for example, believed that too legacy forces are being retained. It declared that "[t]he armed forces' chief planners are holding on to equipment which at best only military museums might be happy to have."

However, the German plan is nevertheless a useful beginning. Germany now has a new and updated concept of war fighting (and peacekeeping) around which to organize its R & D efforts.

* * *

On balance, the emerging trend in Europe appears to be in the direction of embracing, and adapting to the European context, some of the U.S. transformational concepts, including a significant degree of reliance on network-centric warfare and technology as war-fighting discriminators. The European version of transformation also focuses considerably on the shift away from heavy, mechanized armies for local engagement toward expeditionary warfare. When viewed in perspective, the United Kingdom and Sweden are European leaders in these efforts and in utilizing these new concepts in R&D processes, including faster insertion of technology.

In Chapter 3 of this volume, Yves Boyer, Assistant Director, *Fondation pour la Recherche Stratégique*, Paris, France and Chairman of the French Society for Military Study offers something of a contrarian perspective. His essay on *European National Directions on R&D—New Processes and Approaches* argues that from both a strategic and a resource perspective, it is neither wise nor affordable for Europe to imitate the U.S. technology-driven approach to transformation. He highlights that technology is not the only factor in modern warfare, as evidenced by the problems the United States has faced dealing with relatively low tech insurgencies in post-war Iraq. He also argues that transformation in Europe is more focused on shifting toward more mobile and up tempo expeditionary operations, creating professional armies and more modern command structures, and less on technology. Thus, he concludes that Europe should invest mainly in those technologies that support its own strategic goals and operational methods—even if this means diverging from U.S. trends. Mr. Boyer believes European investment strategies should focus on: new sensors and data collection and processing mechanisms to enhance the planning and conducting of military operations; efforts to digitalize the battlefield, unmanned aerial vehicles; theater anti ballistic defense alert systems; space; and efforts to develop capabilities to meet the EU's Headline goals, including such efforts as nuclear, chemical and biological defense, air refueling, and mobile headquarters.

Interestingly, Mr. Boyer's own analysis and examples suggests that even countries such as France—with a different security vision—are nevertheless moving in similar directions to the United States in capabilities acquisition. Mr. Boyer's own review of French programs reveals considerable focus in the French military modernization plan (*Loi de programmation militaire*—LPM 2003-2008) on network-centric warfare and high technology. It includes a focus on digitalization of the battlefield (through, for example, the *Bulle Operationnelle Aeroterrestre* program for ground forces that seeks to link various sensors to shooters) and improved command, control and communications systems. It also includes studies on reconnaissance UAVs and other high technology areas, including sensors and space. The French plan also is like the American one in that, given the large number of existing, fielded legacy systems, France seeks to balance its spending on transformation toward a future 2015 force model with spending on modernization and digitalization of existing force structure. Thus, the French program includes R&D on intelligence gathering systems, including satellites, and a new ELINT/COMINT ship, as well as funding for new and updated traditional war fighting platforms.

In short, a case can be made the key differences between U.S. and French efforts, in addition to size and scope, are more in the realm of overall strategy than capability. The French view, expressed by Mr. Boyer, focuses more on the primacy of developing "transformational" capabilities for the purpose of putting meat on the bones of the EU's Headline Goal objectives than fostering coalition war fighting capability with the United States.

Leveraging European "Dual Use" Technologies for Defense Needs

The downsizing of defense industries and focus on affordability in the United States and Europe has led to a renewed focus on the role of "dual use" technology in defense applications. In essence, there are two sets of related issues. First, to what extent should civilian/commercial integration be encouraged and can it facilitate affordable and innovative defense solutions (an issue relevant in both Europe and the United States)? Second, in the European context of under-investment in defense, can Europe do more to harness its much larger investment in cutting edge commercial technologies—from communications to biotech—for defense applications and utilize these innovative assets to support transformation and close the capabilities gap?

On the first issue, Kenneth Flamm convincingly argues in Chapter 4, in his essay on *Leveraging Dual Use Technologies for Defense Needs*, that "dual use" policy means having the flexibility to leverage technologies and products with both military and commercial applications, including commercializing products developed for military purposes (spin-offs) and adapting commercial products and technologies for military purposes (spin-ons). Certainly, as Flamm explains, "It should be and is a fundamental U.S. defense industrial policy to tap into commercial technology solutions where appropriate. Indeed, given the emphasis on affordability in the 1990s, this longstanding policy became a key feature of U.S. and European armaments policy." Thus, if necessary components or subsystems of defense equipment are commercially available (or can be used with limited modifications), it generally makes no sense to develop military unique products. Generally, military production results in smaller quantities, fewer economies of scale and resulting higher costs—to the detriment of affordability. Thus, the thrust during the 1990s on both sides of the Atlantic was to replace, where possible, military-unique specifications with reliance on commercial off the shelf (COTS) components where possible. This was particularly the case in the arena of information technology and communications, where the establishing defense-unique software is very expensive and resulted in many non-interoperable, standalone systems that were behind the more innovative commercial systems being developed. This takes advantage of the enormous commercial R&D in many technology areas that dwarfs military R&D spending—even in the United States. Simply put, defense R&D for the most part should not duplicate commercial R&D, which generally can perform the tasks at lower cost to the military customer and produce a better result. Moreover, utilizing commercial processes and techniques in the defense arena can put defense firms on a more business footing and lower costs.

A second set of issues relate to European "dual use" policy, and the ability of Europe to leverage off its commercial technology base. In contrast to the defense arena, Europe has industry leaders in telecommunications, information technology and space—key enablers of transformational warfare. Many of the technologies that underlie new Command, Control, Communications, Computers, Intelligence, Surveillance and Reconnaisance (C4ISR) capabilities rely already in Europe today. Indeed, interestingly, Europe has several major firms that have greater "dual use" capabilities than many defense firms in the United States—notably Thales and Ericcson.

Thus, the question is how does Europe leverage these capabilities and potential synergies for defense applications?

First, the existence of these European first-class private sector capabilities highlight that, at the defense systems and subsystems level, the issue is largely the result of the European defense spending gap and not an intellectual or technology gap. In C4ISR, the key is to apply these commercial technologies to defense applications and doing this requires a significant allocation of resources. Yes, Europe can better harness its commercial capabilities but it requires spending to do so.

Second, there are other approaches to harnessing European "dual use" technologies that warrant consideration. These include privatization of governmental laboratories with "dual use" capabilities, the increased role of the European Union and intra-European cooperation, and the role of the European private sector.

The privatization of government laboratories that focus on "dual use" technologies is one available option to leverage these technologies. In this regard, Sir John Chisholm, Chief Executive Officer of QinetiQ Group Plc. of the United Kingdom described at the Symposium how this firm, formerly part of DERA, had successfully leveraged technologies developed for defense applications into the commercial market, and *vice versa*. To be sure, firms like QinetiQ face the challenges of both commercializing military R&D facilities and the becoming profitable in the highly competitive commercial high technology sector. This strategy is by no means risk free and there may be technology areas that are critical to security but nevertheless prove difficult to commercialize; one only has to examine the history of U.S. foundries on radiation hardened parts, for example, to identify such examples. In such cases, the governments face the question of whether to continue to subsidize such technology development for military purposes. Yet, this strategy, not yet emulated on continental Europe, is nevertheless one approach for European governments to consider in order to facilitate "dual use" technology development.

In Europe, national governments have historically played a lead role in technology development—providing the dominant share of public funding for R&D—both individually and cooperatively—with Airbus perhaps the lead example of government-funded programs to develop a European technology base. In Chapter 5, Klaus Becher's essay on *The EU's Role in Leveraging Multiple Use Technologies for Defense Needs* discusses the traditional lead role of national governments in Europe in developing "multiple use" technology—often motivated by a desire to catch up with and compete with the United States—and fragmentation and duplication caused by national primacy in this arena. He also focuses on the increasing role of the European Union in the "dual use" R&D arena. He notes that the existing state of affairs—an uneasy compromise that allocates responsibility for "multiple use" R&D between national governments and the EU Commission (through its five-year programs)—is likely to be perpetuated under EU arrangements still under consideration (including the draft Constitution). In his view, this divided authority—with the EU acting more as a catalyst than a coordinator or leader—creates anomalies and some inefficiencies, and has also prevented the Commission from creating a single, overarching R&D policy for Europe that prevents duplication and fragmentation of effort.

Becher also notes that even the EU efforts on "multiple use" R& D have been primarily civilian in focus with virtually no military component—consistent with traditional limits on the EU's role in defense and security. However, the EU is now gingerly taking steps to leverage "multiple use" technology for military purposes, he writes. Finally, while Mr. Becher suggests that gradually increased EU primacy in "multiple use" R&D is useful (and may contribute to both emerging defense and homeland security solutions), he believes that its ultimate success is intrinsically linked to the evolution of ESDP. "Without an overarching strategic vision and a single defense policy to establish requirements and priorities, cooperative 'dual use' research will continue to lack a focus", he concludes.

One of the areas where "multiple use" technology development has, and will likely to continue to have significant implications, is in space. At the Symposium, Flamm, Becher, and Daniel Hernandez, Director of Research at *Centre National d'Etudes Spatiales* (CNES), the French Space Agency, all noted that space launch, satellites and most but not all of the military and civilian payloads are drawn from the same technological base and are largely developed by the same private firms. Mr. Hernandez' paper presented at the Symposium, entitled "Dual Use Technology in the Space Domain" set forth in Appendix C, provides a useful overview of R&D activities for space applications in Europe. Mr. Hernandez analyzes the longstanding role of "dual use" technology in such military space applications such as communications, surveillance, navigation, early warning, and meteorology. Indeed, given recent European arrangements on space, including the development of a European space policy under EU auspices and the cooperation agreement between the European Union and European Space Agency, space will serve as a test case of Europe's ability to develop space capabilities for both military and civilian use. Two major development programs now underway—the Galileo navigation satellite system and the Global Monitoring for the environment and Security (which focuses on imagery data collection and exploitation) have both civilian and military elements.

Another important element of leveraging of "dual use" technology is the role of the private sector. At the Symposium, Dominque Vernay, the Corporate Technical Director of the Thales Group, noted how several European firms, including Thales, are actually built around "dual use" technologies and the importance of leveraging them for aerospace, defense and informational technology markets. In particular, technology in the areas of software, digitalization, sensors, displays, optronics and microwave are utilized in civilian and military uses. At the same time, Mr. Vernay noted that this leveraging ability is inherently limited and certain unique defense applications sometimes warrant funding for solutions not supported by commercial markets and civilian R&D.

One final and intriguing thought concerning "dual use" technology is the prospect, suggested by Ken Flamm, of expanded Transatlantic cooperation in this arena. While Transatlantic defense R&D cooperation is difficult for a number of reasons (see discussion below), there already is significant commercial cooperation across the Atlantic and a more level playing field (with far less of a disparity in both technological capability and spending than in the military sphere). Indeed, the civilian supplier bases in numerous "dual use" technology areas are already globalized—with networks of Transatlantic partnerships, joint ventures and the like—and fewer restrictions on technology sharing, a key impediment to R&D cooperation. Hence, in areas like space (where there is a fine line between commercial and military applications), communications, software and biotech, among others, there are perhaps better prospects for cooperation in enabling technologies in support of military and homeland security applications. Indeed, the prominence of European capabilities in some areas suggests that such cooperation would be a "win-win" prospect for the United States and Europe.

The Emerging Coalescence of European R&D Cooperation

While there has been a fair degree of cooperation in European defense procurement in recent years (with significant portions of European defense budgets dedicated to cooperative armaments programs), this cooperation has not paralleled in the European R&D community. European R&D efforts remain very fragmented, duplicative, and in some instances in rivalry—with only limited cooperative efforts. Historically, there have been some prior efforts at coordination—most notably through the Western European Armaments Group. However, these efforts have historically been limited in scope and most sensitive activities have been handled on a national basis.

Currently, there is no central European research and defense entity to foster the development of enabling defense technologies (in contrast to the recent emergence of the *Organization Conjoint de Cooperation en Matiere d'Armement* (OCCAR) for joint procurement activities). The lack of historic cooperation reflects European "nations' reluctance to share technical information ...perceived to be of particular military or industrial advantage... or [that] might introduce vulnerability (such as through countermeasures)."[7] Hence, the "more technically self-sufficient Western European nations are reluctant to enter co-operative research in electronic warfare, sensor systems and signature control for fear of undermining their capability lead and industrial advantages in these areas."[8]

The Trend toward Coalescence. Nevertheless, this picture is gradually, and inevitably, changing. In short, there is today a broad consensus in Europe—in governments, industry, and intellectual circles of the need for a shift away from purely national efforts toward greater coordination and collaboration in defense R&D. Indeed, European governments agreed in the spring of 2003 to create an EU Armaments and Research Agency under the auspices of the EU—although many of the details still remain uncertain.[9] Interestingly, there is today virtually no dissent from the need for this overall approach—although there are variations in the motivations for and degree of such collaboration. Indeed, in this regard, it is significant that even the United Kingdom—which until recently had a healthy degree of skepticism about such proposals, has agreed with other EU member states, to the creation of a European Armaments & Research Agency.[10] Somewhat ironically, concerns that the European Commission would create a new armaments bureaucracy in Brussels drove the UK and other

---

[7] Maximizing the benefits of defence equipment co-operation, Report by the Comptroller and Auditor General, UK Ministry of Defence (March 16, 2001), at 30.

[8] MoD Report, supra, at 30.

[9] See Thessaloniki European Council, June 19-20 2003, Presidency Conclusions at 17, and European Council Decision 2003/834/EC (Nov. 17, 2003) (creating a team to prepare for establishment of an agency in the field of defense capabilities, development, research, acquisition and armaments).

[10] This measure also was approved by the European Parliament in April 2003. See Daniel Hannan, "The European Army is on the March" *London Daily Telegraph*, 13 April 2003. According to the Hannan article, this new agency would be responsible for "harmonizing purchasing policy" and "closing the substantial military and technology gaps between Europe and the United States."

nations to develop a separate intergovernmental entity—outside of the Commission auspices.

This shift toward an EU-led defense R&D effort is part of the overall coalescence, documented elsewhere, of a European defense identity under EU auspices. With the development of the European Security and Defense Policy and the corresponding EU Headline Goals for capability development, it is not surprising that Europe is moving toward better coordinated defense R&D efforts. Indeed, there have already been a number of nascent moves in this direction that warrant serious consideration:

- *The Letter of Intent (LOI) Framework Agreement.* This agreement, signed by Europe's six leading armaments producers, requires more exchange of information on national research and technology strategies and programs, and seeks to foster more cooperation and competition. It sets forth principles for collaboration, including the elimination of project specific-jus retour requirements.

- *The EUROPA Memorandum of Understanding.* This agreement, executed by members of the Western European Armaments Organization, is designed to translate the LOI principles into a practical basis for cooperation; the Memorandum provides greater transparency in the exchange of national research and technology plans while permitting nations to conduct cooperative research in smaller and more manageable groups if they so chose.

- *The European Technology Acquisition Program (ETAP).* From a practical standpoint, perhaps the most important development is the announcement by six European countries (the UK, France, Germany, Italy, Spain and Sweden) to launch ETAP, an umbrella effort to develop stealth, advanced avionics, vehicle systems and weapons integration, and other technologies applicable to air warfare). While ETAP is significant in that it does reflect a commitment to work together on research technologies in a key sector, many questions remain about the program, including whether it will focus on manned or unmanned future aircraft, for combat, command and control, surveillance or other uses, and which specific technologies it will seek to develop.

Why Coalescence? This clear coalescence toward more coordinated European defense R&D efforts reflects a range of mixed motivations. Some in Europe, including national armaments directors, view this effort in economic terms, as a logical step to promote affordability and innovation in defense R&D and avoid the duplication and fragmentation of the past. Moreover, this helps to allow the "demand" side of defense markets, which has remained largely balkanized, to catch up with supply side developments, where consolidation has create a European defense industry as distinct from national defense industries. Not surprisingly, industrial support for an EU Armaments & Research Agency is strong. At the recent R&D Symposium, Daniel Deviller, the Senior Vice President for Industrial Research & Technology and Chief Technology Officer at the European Aeronautical, Defense and Space Company (EADS) noted the paradox faced by European defense firms. On the one hand, investment in R&D is a driver of defense innovation and commercial success in the marketplace. On the other hand, the lack of European R&D collaboration is a lag on European industrial development. In this regard, he noted that defense R&D programs in European countries are chronically under-funded and thus unable to achieve critical mass. While he viewed a number of the newer European R&D cooperation mechanisms as (EUROPA, ETAP, etc.) as having potential, none is comprehensive and each one serves to carve out independent "niches" of cooperation; as a result, fragmentation is likely to remain for some time.

Additionally, the shift toward transformation creates yet further powerful incentives for European R&D cooperation. Simply put, the focus on network-centric warfare and technology as battlefield discriminators requires R&D investments that national governments in Europe cannot afford on a stand-alone basis. The types of C4ISR architectures that would enable this type of warfare require substantial outlays that European nations cannot make alone. Indeed, to not cooperate would mean that Europe would fall even further behind the United States in this critical area.

Others, however, within Europe base the drive for European R&D cooperation on geopolitical considerations, including a natural extension of ESDP and counterweight against U.S. R&D supremacy. For example, Yves Boyer argues, in his essay, that Europe must work within itself to develop a competing technological base and a single indigenous research organization in order to meet its own strategic objectives. Indeed, some argue that the successful U.S. operation in Iraq has accelerated the drive for European cooperation in defense and security, including in the armaments arena.

Open Questions on European Defense R&D. While the trend toward increased European defense R&D cooperation is now clear, numerous questions remain. First and foremost, as discussed above, defense R&D is derivative of an overall concept of security and war fighting. Yet, today Europe lacks an overall strategic framework for acquiring military capabilities beyond its focus on low intensity Petersburg principles. Simply put, there is no "European" assessment of threats, evaluation of military capabilities needed to address such threats (typically in the form of requirements), and prioritization of what R&D efforts would be most useful in developing capabilities that meet such requirements. While various European nations are moving toward military transformation, there is no European conception that has been embraced. Hence, developing this set of "first principles" undoubtedly will and should be the prime focus of any new European Armaments entity.

Second, there are serious questions of the scope, degree and pace of European defense R&D cooperation. Will the new EU entity simply catalyze and coordinate R&D or seek to centralize it? And what will the time frame for these efforts be? Clearly, creating the Agency outside of the Commission was intended to limit its authority and retain national sovereignty in this arena. While some in Europe believe a more centralized approach—a single agency leading all R&D (much like the U.S. DoD has today), political realities in Europe precluded moving any further in this direction. Finally, there are question whether the evolving frameworks for European defense R&D cooperation, from the LOI Framework Agreement, to ETAP to a nascent European Armaments and Research Agency, by their nature, exclude cooperation with the United States or participation by U.S. industry.

In Chapter 6, Andrew James's addresses many of these issues in his essay on *European Defense Research & Technology (R&T) Cooperation—A Work in Progress*. First, from a macro perspective, Mr. James sees little prospect for increases in European defense R&D spending, but believes that maintenance of current budget levels will lead to better spending as resource constraints foster deeper cooperation. However, he sounds an urgent note that Europe's chronic under-investment in defense R&D has caused it to fall significantly behind the United States in critical mission capabilities. Thus, unless research spending is increased and employed more effectively, Europe will be unable to overcome its technology deficit and will be consigned to a permanently inferior position relative to the United States.

On balance, Mr. James' bottom line is that greater European collaboration in R&D is inevitable given the press of operational requirements and the paucity of individual national budgets but will not likely lead to a "step change" in European R&T capabilities. Instead, he foresees a more incremental and hesitant move towards consolidation of R&T efforts under a single European research agency, gradual movement towards sharing of technology data, and the maintenance of national programs for some time—despite the consequent duplication and fragmentation of effort. His sobering assessment reflects a number of factors. First, European governments tend to favor ad hoc arrangements outside of the EU, mainly for reasons of sovereignty and industrial security. Second, he points to history—including the limited past efforts and more recent efforts at collaboration. His cautious optimism reflects the fact that such recent steps as the LOI Framework, the EUROPA MoU, and the European Technology Acquisition Program (ETAP) have yet to produce significant results.

Transatlantic R&D Cooperation

Another approach to enhancing the effectiveness of defense R&D spending on both sides of the Atlantic would be a greater degree of cooperation between the U.S. and its European allies on transformational research and development programs—both "top down" and "bottom up." As used herein, "top-down" refers to government cooperative programs while "bottom-up" referred to industry-driven activities.

The Benefits of Cooperation for Transformation & Coalition Warfare. Many observers have called for such cooperation, recognizing that it would eliminate duplication of effort, broaden the technology base upon which programs could draw, and provide for greater sharing of risk. For Europe, such cooperation could help close the existing "capabilities" gap and fuel its transformational efforts—especially in the C4ISR arena, where the United States has an enormous ongoing

R&D effort. Moreover, such cooperation could potentially enhance coalition war fighting capabilities and promote interoperability between forces operating at different capability levels.

Interestingly, there are mixed European views today on both the merits of and realistic prospects for closer Transatlantic R&D cooperation. On the one hand, a number of leadesr in the European armaments community, including Mr. Gould in his essay, stress the desirability, if not the absolute necessity, of closer R&D collaboration with the United States to promote interoperability. However, Mr. Yves Boyer argues in his essay, and other Europeans agree, that closer Transatlantic cooperation will only become possible when Europe is capable of participating as an equal with the United States. Otherwise, they argue, Europe will find its own technology base languishing behind that of the United States. Further, other Europeans, including some in industry, believe that increased Transatlantic cooperation will only become attractive to the United States when Europe improves its own technology portfolio. Andrew James notes in his essay that without increased European cooperation, there is relatively little that Europe has to offer the United States except in certain narrow niche technologies, which in turn does not create incentives for the United States. to participate in cooperative programs with Europe. Paradoxically, James points out that if Europe could gain greater access to U.S. R&D programs, this in turn would strengthen the European technology base—making continued cooperation with the United States more attractive to both sides.

Cooperation Realities: Yesterday, Today & Tomorrow. Yet, the reality is that there has been limited Transatlantic R&D cooperation in recent years—since the gradual phase out of most of the Nunn coalition warfare programs in the early 1990s—and few prospects of a broadened engagement in the near term. While there has been a significant amount of dialogue on cooperative R&D programs, resulting in more than one hundred Memoranda of Understanding (MOUs), and bi-lateral and multi-lateral meetings at many levels, little has actually been accomplished by way of real development.

Perhaps the most important efforts in this arena was the formation in 1996 and operation since that time of the International Cooperative Opportunities Group (ICOG), a multi-lateral group sponsored by the so-called "five power" national armaments directors (France,

Germany, Italy, United Kingdom, and the United States). At the recent Symposium, Dr. Spiros Pallas, a former Director of Defense Research and Engineering (DDR&E) in the U.S. DoD, explained the principles and processes through which the ICOG operates and other vehicles for defense research cooperation. The ICOGs were tasked to identify programs for potential cooperation based on factors that create a successful cooperative program: 1) the degree of requirements commonality; 2) the extent to which the technologies, strategies, and budgets of the potential partners are complementary; the potential for international industrial teaming; and 3) the perceived benefits and risks associated with execution of an international program. To date, however, Dr. Pallas noted that this approach has not yielded significant results. Some of the ideas developed in the ICOG process actually resulted in European cooperative efforts while others undertaken Transatlantically have not produced significant success stories.

NATO itself also has limited R&D activities. The leading NATO entity in this area is the NATO Research and Technology Organization (RTO), a voluntary membership organization that serves as a facilitator for the creation of cooperative multi-lateral programs and exchange of ideas. Other NATO research organizations include the NATO Command, Control and Consultation Agency. The work of the RTO has resulted in modest successes in such diverse areas as urban warfare, chemical and biological defense, hybrid electric vehicles, federated simulation and training, and charge-coupled devices.

Finally, the recently signed Declarations of Principles for Defense Equipment and Industrial Cooperation between the United States and a number of European allies (the United Kingdom, Italy, the Netherlands, Norway, and Spain) also calls for more harmonization and information exchange concerning research and development programs, avoidance of unnecessary duplication, and increased cooperation. Yet, the reality is that these aspirations have not yet translated into practice either.

Today, the United States continues to express openness to foreign participation in its R&D programs. Indeed, the United States does have a series of modalities designed to identify and evaluate interesting foreign technologies and defense solutions. At the Symposium, Dr. Anthony Tether, Director, Defense Advanced Research Project Agency (DARPA), highlighted the potential vehicles for international participation in DARPA programs, particularly the role to be played by Transatlantic joint ventures and strategic partnerships. Yet, this openness has not been reflected in practice and few of these solutions have been integrated into U.S. forces.

Table 1. Transatlantic Defense R&D Cooperation

Program Name	Participants	FY 2000 Actual	FY 2001 Actual	FY 2002 Actual
AGM-88 HARM Upgrade	GE, IT	$36,773	$39,409	$13,630
AV-8B Harrier II Plus	UK (Engine)	$1,572,297	$36,410	$32,897
Directed IR Countermeasures		$8,864	$7,465	$1,706
IRC Gas Turbine	UK, FR	$25,685	$9,547	$3,921
Joint Strike Fighter	IT, NO, DN, NE, CA, UK	$522,009	$789,511	$1,747,739
JSTARS		$700	$3,240	$0
MEADS	GE, IT	$0	$0	$73,645
MIDS (DoD)		$29,336	$16,100	$7,601
MIDS (Navy)		$42,706	$4,138	$1,309
MLRS Improvements	GE	$64,749	$68,886	$111,389
Standard Missile II	IT, FR, GR, NE	$625	$1,183	$1,309
Advanced Concept Technology Demonstrations				
C4I for Coalition Warfare		$1,200	$2,000	$1,200
CinC 21		$2,200	$9,900	$13,600
Coalition Aerial Surveillance		$2,400	$1,900	$2,500
Coalition Theater Logistics		$0	$1,500	$3,200
		$2,309,544	$991,189	$2,015,646

Source: U.S. Department of Defense, FY 2003 Budget Submission

In short, the bottom line reality is that there have been few new significant Transatlantic R&D starts during the last decade—with the Joint Strike Fighter (JSF), Medium Range Air Defense System (MEADS), and the Multifunctional Information Distribution System (MIDS) being the exceptions to the rule. Most other cooperative projects have been very small. Moreover, there has been little if any significant foreign participation in U.S. transformational programs—from the U.S. Army's Future Combat System to any of our major C4ISR projects. Indeed, as Table 1 shows, U.S. cooperative R&D amounts to little more than 0.5% of the total U.S. R&D budget, and JSF accounts for the lion's share of that.

It should be recognized that considerably more cooperative R&D is taking place at the industrial level, either between companies on either side of the Atlantic, or between business units of companies that now have a Transatlantic identity. These efforts tend to be more program-oriented, however, and focus on application of existing technology to specific requirements.

Impediments to Transatlantic Cooperation. In practice, the limited nature of Transatlantic R&D cooperation—both top down and bottom up—is in part a function of well-known impediments. These include not only requirements harmonization and budgets, but also very significantly export control and technology transfer restrictions. Simply put, the United States continues to maintain a significant disconnect between national armaments policy and our technology transfer policy. The U.S. tends to open the door to engagement on armaments—missile defense, the NATO Airborne Ground Sensor, and JSF are but the most recent initiatives—but the U.S. does not follow through with a coherent technology transfer policy to support these key programs. In essence, this "disconnect" reflects that the old "paradigm"—that U.S. national security derives from our technology and industrial leadership in defense technology—is still alive and well at the Pentagon, and affects U.S. decision-making on technology release and disclosure to even close allies.

Moreover, the history of the cooperative efforts in which the United States has participated have, over the years, been very mixed—with many of the same problems that plagued intra-European cooperative programs. While there have been successful cooperative programs (for example, the Harrier and the RAM programs), there also have been a long history of problems highlighted in numerous studies over the years and a recognition that many of the theoretical benefits of international cooperation (political, economic, and operational) are difficult in practice to achieve.[11]

---

[11] See, e.g., Birkler, J., Lorell, M., and Rich, M., "Formulating Strategies for International Collaboration in Developing and Producing Defense Systems," RAND Issue Paper (1997) ("Despite a long record of international procurement collaboration among European partners and between the United States and its allies, the outcomes of past programs have been, at best, rather mixed. Attaining many, if not most of the potential economic, operational, and political benefits that theoretically should flow from joint R&D and production programs has proven difficult. . . .")

Finally, the reality is that the relative degree of Transatlantic R&D cooperation also is a function of broader geopolitical and war fighting realities. Thus, the greater the shared Transatlantic security interests and objectives and the stronger the commitment to coalition war fighting, the greater degree of R&D cooperation can be expected. Today's relatively limited amount of cooperative engagement in defense R&D thus reflects underlying, and long-term, divergences in these areas. The reality is, however, over the long-term, mutual interests on both sides of the Atlantic, including the growing consensus on defense transformation and the importance of technology in war fighting, will likely fuel greater R& D cooperation; economics, politics, and the need for technological innovation will likely drive the move toward enhanced cooperation.

Perhaps the most useful area for shared R&D efforts in the short term relates to the development of the NATO Response Force (NRF). This effort to stand up a NATO high intensity expeditionary force and incentivize European capability acquisition offers interesting potential opportunities for collaboration—especially in the C4 area vital to transformation and improved interoperability. In a sense, the standing up of the NRF and identification of necessary improvements for it to function—how can our forces be made interoperable and have access to the same battlefield picture—serve as a test bed around which cooperative R&D can be organized.

In Chapter 7, Dr. Masako Ikegami provides us with a comparative analysis in her essay on *International Defense R&D Cooperation: From Competition to True Cooperation—The Case of US-Japan Defense R&D Cooperation in Transition*. Specifically, she evaluates the evolving cooperative defense relationship between the United States and Japan, which she argues offers some useful lessons for Transatlantic defense R&D cooperation. Tracing the history of U.S-Japanese defense cooperation, she identifies some of the peculiarities and constraints affecting this relationship, including the Japanese constitution, arms export restrictions, and protectionism. Against this, she posits two "meta" trends in defense R&D: the commercialization of defense technology and the concomitant globalization of defense R&D. These two trends and the close commercial relationship between Japanese and American technology companies, is creating greater bottom-up U.S.-Japanese R&D cooperation in military applications. Dr. Ikegami notes how commercial technology in critical areas such as information technology, electronics and telecommunications is beginning to affect the development of new systems and the structure of the defense industry. In her view, impediments to cooperation still remain—in the form of U.S. technology transfer requirements and Japanese protectionist industrial policy. Yet, in the long-term, she argues that cooperation in defense R&D has become an imperative for all countries—even the United States. The cost of developing new weapons systems and capabilities, the need to meet a wide range of diffuse threats, the difficulty of mustering the political will for the large commitments of resources needed to be self-sufficient in R&D is rapidly exceeding the capacity of even the last superpower.

The Challenges for the Future

In sum, as this volume discusses in depth, the issues surrounding the development of European defense R&D are complex and cannot be understood in a vacuum; they are in fact a function of economics, geopolitics, and technology. In essence, however, the analysis suggests that Europe faces five principal challenges in the defense R&D arena. These challenges bear consideration as the reader examines each of the following chapters:

- *First and foremost, organizing principles for European R&D spending.* Europe needs to develop a common strategic vision: the roles and missions will Europe undertake and how will it fight. This in turn will determine where R&D spending should go. Without a strategic vision, there is no ability to prioritize, and centrifugal forces will cause national programs to go off in different directions. Europe will need to determine its emerging transformational concepts and the nature and degree of its emerging thrust (now only on a national basis) toward expeditionary, and network-centric warfare and greater reliance on technology in warfare.

- *Second, new processes are needed* at the national level to improve efficiency of R&D spending, including greater freedom for entrepreneurial ventures, possibly involving Privately Financed Initiatives (PFIs) and public/private partnerships, and a role for universities—not much emphasized in continental Europe. Commercial companies must be brought into the defense market in order to leverage "dual use" technologies, in which Europeans have real competitive potential.

- *Third, how to ensure the type of environment* to foster innovation in defense and "dual use" technologies is a key issue in Europe. Europe must learn how to use competition to foster innovation even within a consolidated defense market with few buyers and even fewer producers. While competition is not the only driver of innovation, it is a major one, and the ability of European companies to compete globally in key commercial fields such as telecommunications points to the benefits of technological competition in a relatively unregulated market.

- *Fourth, the coalescence of European defense* and the need to enhance European cooperation through development of a European Defense R&D agency. There is no doubt that this will happen, but the form, the timing, and at what cost are all crucial variables. Can existing mechanisms be leveraged to form a new entity without creating runaway bureaucracy? Previous efforts in this direction (e.g., Eureka) failed due to lack of common focus and accountability. Future initiatives need to demonstrate results to gain additional adherents. Closely related to this is the question of whether Europe can create a "European" defense market and buyer without creating an autarkic "Fortress Europe" that closes the market to the United States—to its own detriment from the standpoint of competition, innovation and war fighting capability.

- *Finally, there is the challenge of improving transatlantic cooperation and addressing the impediments to such cooperation*—from the standpoint of arming and fighting ourselves together, but also from the geopolitical standpoint of cementing our alliance relations. While there are numerous and complex issues here, the reality is that U.S. rules, policies and attitudes governing the release of technologies to our European allies continue to be a central stumbling block to enhanced cooperation.

Chapter 2
European National Directions on R&D United Kingdom R&D Policy and Transatlantic Cooperation

David Gould

Why is defense R&T so important? The United Kingdom's New Industrial Policy (October 2002) gives two reasons:

- "Military advantage is generally enjoyed by all nations with access to high technology"

- "[I]nvestment in research and high technology is the critical factor in the future prosperity of the defense industry"

These are mere aphorisms, but like most aphorisms, they hold an element of truth. If the United Kingdom is to continue to enjoy access to high technology (and the military advantage that accrues thereby), and if investment in R&T is the key to the continued prosperity of the British defense industry, then it is clear that the UK cannot continue managing defense R&T in the same way it has for the last thirty years (and in fact probably more than that). There are several underlying reasons for this, but they all revolve around the idea of scarcity. All countries face problems of scarcity with research and technology. Even if money is plentiful, defense is in competition for R&T with the private sector. If one wants to deploy technology into the armed forces, one must think about training, performance, prioritization, and cost-benefit ratios. In the end, the problem is indeed managing scarcity in some way or another. In that case, the first question that must be asked is "Which technology is going to provide the most benefit in the given time frame?" Traditionally, this has been done through operations analysis (OA). Everyone familiar with OA knows that it is only as good as the questions asked in the first place. While OA works quite well in a (relatively) static environment (which pertained in the Cold War) for testing concepts and solving problems, when one is facing rapidly evolving, asymmetric threats, OA does not provide a good indication of what to do with technology early in the development cycle. The new security environment requires that we do something new.

The UK View of Research—The Old Approach

Even in the new security environment, some of the old approaches are still acceptable, at least in some areas. For instance, in jet engine technology, civil and military requirements do not always coincide, but they share common fundamentals and material. They tend to diverge only where civil and military operational imperatives evolve in different directions. Despite some problems over the last three or four years, the United Kingdom has had a steady relationship with Rolls Royce and other corporations in joint government and private sector funding of military engine technology, leading to progressive benefit. The cutting edge in this area today is the Joint Strike Fighter (JSF) Short Takeoff/Vertical Landing (STOVL) engine, the specific British contribution being the lift fan and the rear of the engine. Producing as much thrust as the Tornado's RB199 turbofan engine when it is not in reheat (afterburner), the lift fan is completely dry (i.e., there is no fuel injection or combustion chamber), just a driveshaft linking the turbine to the fan. The JSF lift fan shows that one can develop, produce and then cooperatively put into the U.S. market a product that is both excellent and a real advance in technology.

Now the U.S. Air Force has a research program called Versatile Advanced Affordable Turbine Engines (VAATE), which ironically is being run by Allison Technical Company (ATC), which is owned by Rolls Royce. But when Rolls Royce and ATC went to the U.S. State Department to propose they might actually work together on this program, they were turned away and were told that it was not for the private sector to propose cross-government research into defense technology. This incident points up some of the shortcomings of the traditional way of managing defense R&T, and the need to find better, more flexible avenues of cooperation.

But in some areas, there is no commercial sector on which one can fall back, in which case the government must focus its funding on specific technologies. In such areas, consistency of investment pays dividends. For example, the UK has been working on nuclear biological and chemical (NBC) protection technology, prophylaxis and medication since poison gas was first used on British forces in 1915. And we have actually found a way of managing that technology very effectively through research, through the setting of requirements, and through working with the private sector and particularly with France and US. This approach has worked very well over the years, and continues to work well. This work really needs to be done by government, because the opportunity for carrying out clinical research on people who have been exposed to nerve agents is actually quite small.

New Approaches

Thus, while the traditional approach to managing R&T still has some applicability now and in the future, the UK is convinced that the future environment in general will require new approaches that provide greater responsiveness, flexibility and efficiency.

The UK is presently focusing on three new approaches to defense R&T:

- ITAPs, which plan for technology insertion at the earliest opportunity over the life cycle of developmental and fielded systems by managing capability and technology *in parallel*, not sequentially
- New Science & Technology Strategy, defined as a set of outputs which deliver capabilities
- Rapid Test & Evaluation of Concepts and Prototypes, in which flexible facilities are used to model the impact of technology and ideas in the battlespace

In Integrated Technology Acquisition Plans, technology and concept development are linked. As shown in **Figure 1,** the ITAP process begins with a broad range of operational concepts and broad range of applicable technologies. These must be narrowed down and managed so that applicable technologies mature as the concepts develop. By the time the concept and technology are pulled through into the capability for the armed forces, the two are coming through in parallel. ITAPs have a number of beneficial aspects. They provide for auditable identification of critical technology requirements early in the development cycle, and they provide complete coverage of the technology space. This allows for analysis of tradeoffs between Key User Requirements (KURs) and various concepts, timescales and cost options. Early identification of critical long-lead technologies can ensure adequate and consistent technology investments keyed to concept development timelines, as well as providing links into the UK's Applied Research Program, private venture research, etc. Properly implemented, the ITAP process can, in theory. provide the confidence needed at major program decision points that sufficient technological maturation can be achieved and that KURs and the transition from the technology development to the demonstration phase, can be met with manageable risk.

One area on which the UK is particularly focused at present is the Future Offensive Air System, (FOAS), which is the Tornado GR.4 replacement. Traditionally, this program would begin with question about the platform: Is it an airplane, or not an airplane. Under the ITAP process, this is the wrong question. Instead, one begins with a definition of the capabilities and effects one wishes to achieve, and works back into the enabling technologies, and then forward into operational concepts and capabilities. In the case of FOAS, there may be a manned aircraft or a manned aircraft may be just one of several platforms delivering the capability; either way, FOAS is much broader than simply the replacement of one aircraft for another.

Figure 1. Integrated Technology Acquisition Plan Process Flow

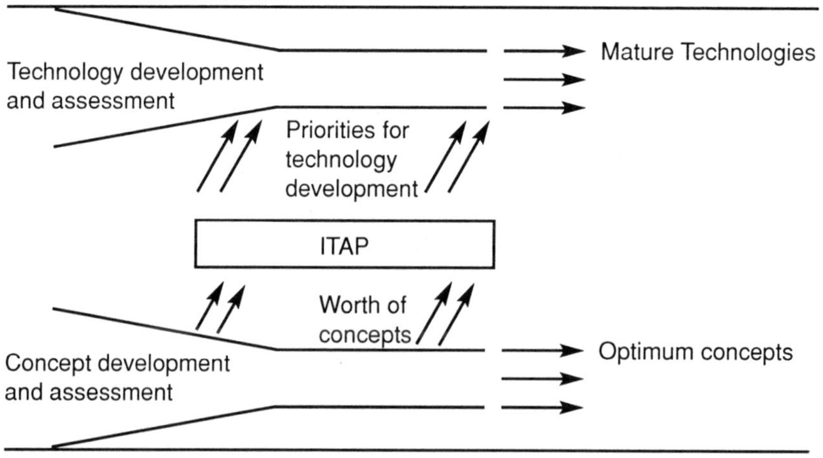

The other main ITAP focus at this time is the Future Rapid Effects System (FDS), the British Army's plan for transformation from heavy armor vehicles with heavy protection into a light deployable formations. It is not a vehicle program. Rather, FDS is a range of capabilities that will enable the army to achieve its objectives. Thus, we do not begin by saying that one is trying to build a new vehicle. Rather, we investigate the full range of enabling technologies up front before deciding which approaches will deliver the most capability in the given timeframe within the defined resource constraints. In short, we, finalize system specifications late, but choose the key technologies early in the cycle, mature them and pull them through when we can. Technology development, or pull-through, is matched to the acquisition program, minimizing two factors that contribute most to project failure—cost and time overrun.

Examining the projects managed by the Defense Procurement Agency ("DPA"), with the exception of one or two spectacular failures like Nimrod AEW, one finds that most do work and meet requirements, but they tend to arrive well behind schedule and over budget. And according to DPA's own analysis, some seventy percent of the time, the reason for schedule and cost overruns is poor technology management. The ITAP process is intended to ensure that technology is developed in such a manner that it becomes available to programs when the capabilities it provides are needed.

The second pillar of the UK's new approach to defense R&T is the New Technology Strategy, which presents a completely different way of looking at the research and technology budget, as shown in **Figure 2**. All military forces like to measure their R&T budgets, but how many people, even in the defense ministry, actually look at the output from that budget and decide to actually allocate that budget on the basis of the value we expect to give output.

Thus, the R&T budget is divided into seven sectors, or "outputs," each of which has an "owner" (an agency responsible for generating the output) and a "stakeholder group" (users of the capabilities provided by the output). Each output owner will be allocated resources, and empowered to deliver the output to the stakeholders. More importantly, the owner will be accountable for the delivery of capabilities, and future allocation of resources will be determined by the effectiveness with which the owners are able to deliver.

Figure 2. Output-Based Technology Strategy

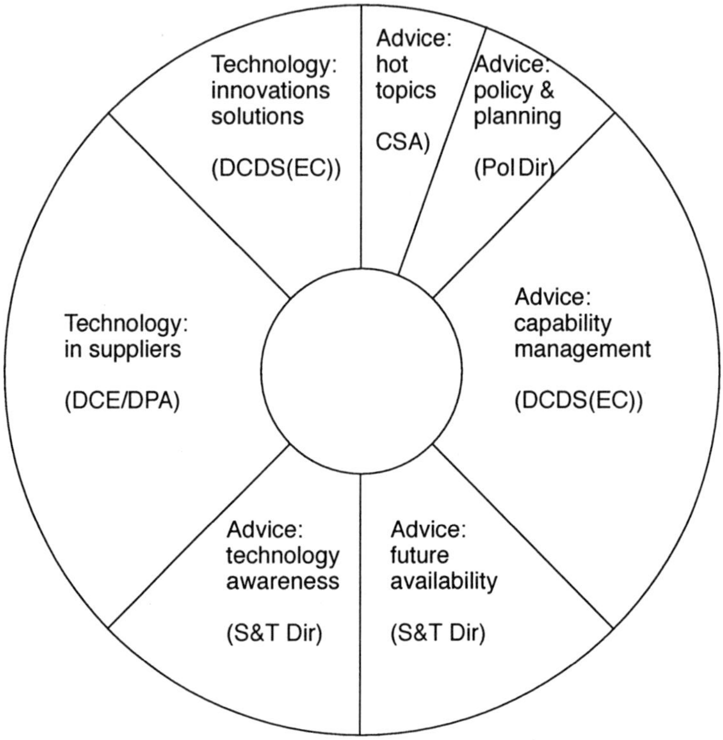

```
                 Advice:
   Technology:    hot      Advice:
   innovations   topics    policy &
   solutions              planning
                  (CSA)
   (DCDS(EC))              (Pol Dir)

                          Advice:
   Technology:            capability
   in suppliers           management

   (DCE/DPA)              (DCDS(EC))

          Advice:      Advice:
          technology   future
          awareness    availability

          (S&T Dir)    (S&T Dir)
```

I find one of the interesting things on this chart is how many of seven outputs at present provide knowledge and advice, rather than providing capability. This will have to change in the future, shifting more towards putting technology in the supplier base and looking at technically innovative solutions. This will be quite a cultural challenge for the traditional scientific community, who are not used in the military and defense in the UK to looking at their work in this way. The idea that they should take some of "their" money, and give it to someone who acquires equipment to do things with, is something of a paradigm shift.

My responsibilities at DPA include "Technology in the Supplier Base." Under the theory of the new S&T strategy, I will be given a budget, and it will be my job to undertake certain initiatives to produce certain outputs. Ultimately, I must report to the Science and Technology board on how successful I have been in doing that. Based upon their relative success, resources will be reallocated among the different outputs in the next budget cycle.

The key to success in this area will be to break down the wall between that which is R&T and managed as R&T, and that which is acquisition and managed as acquisition, creating a continuum not into the old cycle doing research, stopping to think about it, working out some new concepts and starting over doing something different. Once the process is recast as a continuum, one can do more with the supplier community by way of demonstrations, and the pull through of technology by themselves.

To be more specific, under this particular output, directed research and technology demonstrations will be undertaken by industry, while the DPA works with industry to sustain the supplier base. This will enable technology development in industry to proceed at a pace and direction that would otherwise not have been taken. It will include the promotion of technologies derived from government-to-government agreements, and the sponsored exchange of personnel between industries and governments. Finally, the DPA will encourage investment by overseas concerns in support of the Defense Industrial Policy as a means of broadening and deepening the UK's defense technology base.

The third pillar of the new approach, Rapid Test and Evaluation, was driven by Urgent Operational Requirements (UORs). Whenever the military is deployed on operations, one discovers it is never the operation anticipated, so one must do a number of things to adapt your equipment to that new situation. In the UK, these are addressed by UORs, and the DPA must pull ideas through, test them, evaluate them and do it very quickly.

This was done during the recent war in Iraq, with some 500 million pounds worth of acquisition from the DPA. All material arrived on time, despite the fact that we were not allowed to start early or allowed to do things in a conspicuous way (officially, it hadn't been decided if anything was going to be done at all). Despite this, a very great deal was accomplished in a very short period of time.

If that works so well as an operational stimulus, why on earth can't it be done all the time— as the normal way of doing business? Rapid

Test and Evaluation is DPA's response to that challenge. For example, under the JUEP, the DPA took some new UAVs and put them in the hands of some operators to evaluate their capabilities and identify elements that need correction and modification. This gives the operators some initial capability, which can be converted into a more systematic project.[1]

The approach being used in the Network Integration Test & Experimentation (NITEworks) is considerably more complex and was created to model the UK's Network Enabled Capability (NEC). Initially known as the Experimentation Network Integration Facility (ENIF), it is a federated facility linking a number of reference sites, test sites, integration facilities, organizations within QinetiQ, the School of Signals at Blandford, General Dynamic UK (which is working on the Bowman communications system), the dstl government laboratory (which is very good at validating simulations and ensuring that they reflect the real world). The federated approach is designed to create a synthetic environment in which to test the network and make it work.

In a system of this complexity, one cannot do everything at once, nor can one afford to try. It is therefore essential to do the right things first and move on. This gives you the ability to make changes in how the network works and evaluate the results. Does a specific change actually produce the desired output? If it does, it can be turned into a project and the technology inserted into it. Experimentation may not even involve technology; it may simply involve changing the way in which decisions are made. It might just be behavioral, but it provides a way to actually simulate the real world and indeed, to evolve the "real world" into the system simulated in the virtual environment.. This requires some innovation on the commercial front as well as on the technical front. It requires innovation from the authorities who approve the project, because they would like to know that there is tangible product at the end that does something and can be seen and touched. Instead, they must learn to rely on some very clever people who do "simulations and things" which makes description of the output problematic.

---

[1] This is not unlike the Advanced Concept Technology Demonstration program employed by the U.S. Department of Defense.

BAE Systems and QinetiQ are co-primes with DPA because they have the best understanding of the network. They, in turn, are pulling in a wide range of other companies, which does actually challenge the commercial community, because it means sharing data, sharing technology, and sharing IPR. There is also the problem of working out how project workshare (which is where the bulk of the commercial value lies), so that all industrial participants feel they have a fair chance of winning some business.

DPA relies on BAE Systems as a prime contractor leading an inclusive industry partnership; this allows the MoD to access key industrial skills, IP, and facilities. In regard to NITEworks, BAE is working with QinetiQ, AMS, Thales, GDUK and Logica CMG to design an intellectual properties framework, without which deeper industrial participation would be difficult to attain. The MoD also has ongoing discussions with more than forty companies concerning MoD/Industry partnerships, and it is developing a Small/Medium Enterprise Engagement Strategy to bring in the kinds of entrepreneurial startup companies that provide cutting edge technological innovation.

This approach in particular is part of the answer to the question posed earlier: "How does one change from the old culture of deciding what you need to do, into a world that is more attuned to asymmetries and rapid change."

Conclusion

In sum, scarcity demands collaboration and specialization, resulting in more effective use of resources. Achieving this means changing behavior, whether it is between or within governments, It is not just enough to say "We need to do better in Europe" and "We need to write more MOUs." This is not going to do the trick at all. The UK, and Europe more generally, needs focus, needs technology insertion planned early, methods for deciding on the right technologies to pull through, and methods for maturing them and ensuring that they mature in line with acquisition program. One needs measurable outputs for technology development, together with good test and evaluation, that allows assumptions to be tested in the real world or at least in the simulation of the real world.

Chapter 3

European National Directions on R&D—New Processes and Approaches

Yves Boyer

Today, international life oscillates between apparent stability and anarchy. Such volatility has many causes: "a return to history" with the traditional dynamics of interstate relations; new international issues that have not yet found satisfactorily solutions (environment, pandemics, etc.); and flagrant disorders and injustice that contribute to disturb the current world order. The quest for an ideal world order seems so utopian that apparently, as a practical solution, only one choice would be realistically plausible, as argued by the so-called "*European foreign policy establishment*" in a Financial Times editorial: "a new covenant between the United States and Europe in order to perpetuate Western supremacy." What could it mean?

Together the democracies of the West have a moral duty—as well as long-term interest—to guarantee stability at the world level in order to let the market economy bring prosperity and development. In order to do so, the United States and Europe should unite their strength. The imbalance in military capacities between the two sides of the Atlantic would make it seem "natural" that Washington exert a leadership role tempered by a capacity for influence collectively played by the EU or by individual nation state, the classic example being Great Britain.

From a military standpoint, the homogeneity of the alliance would have to be re-established according to U.S. military thinking. This is the sense of the present attempt by the Bush administration to "transform" the allied military forces according to principles and mechanisms defined by the U.S. military. The transplant of American understanding of warfare to U.S. allies in Europe is now the task of the new NATO "transformation command" (Allied Command Transformation, or "ACT"), a direct counterpart of the United States with the Joint Forces Command (JFC). Both ACT and JFC have the same location at Newport and same U.S. commander-in-chief, Admiral Giambastiani. In this framework, the EU is subordinate to the United States with but a slim power of influence, and the weapon systems it develops will be strongly influenced by U.S. military doctrine. Military R&D spending would then be rationalized under the pretext of Alliance efficiency. A certain number of countries in Europe might well accept such a situation; others certainly would not.

Indeed, such a scenario raises a lot of questions. The first one relates to the nature of current U.S. foreign policy. Are the radical changes decided by the Bush administration transforming U.S. foreign and security policy from being a world policeman to become a kind of new "conqueror of the world," a transitory or a permanent feature of U.S. foreign policy? Senator Fullbright, in his book written in 1966, *The Arrogance of Power*, already warned his fellow citizens not to succumb to the temptation of moving from an imperial policy to an imperialist policy. Encompassing many facets, such a policy is, however, not exclusively grounded on an overemphasis on military affairs.

However, in the security field, the United States has accelerated its advance at the world level notably by injecting huge sums of money in procurement and in military R&D. Such an advance puts into doubt the possibility, in the future, of allied forces co-operating with U.S. forces during a military intervention. Accordingly, it raises the question for Europeans of the logic of imitating the path followed by America in the domain of military R&D.

The U.S. Advance in Military Affairs

The roots of what provoked the current transformation in military affair hark back to the 1960's. Among the causes of the present changes, two are noteworthy. The first is related to the expansion of U.S. nuclear arsenals including the need to build complex command centers to control strategic forces, as well as the competition between Washington and Moscow in space programs. Both phenomena accelerated the quest for immediate and dramatic improvements in new technologies (e.g., new materials, new communication systems, increased role for computers, new propulsion systems, etc.). As a result of that competition, conventional weapon systems were also dramatically improved, particularly in terms of command, control, and precision. Secondly, the dynamic of the confrontation between NATO and the Warsaw Pact accelerated the transfer of high technology from nuclear and space activities to conventional weapon systems. Gradually, new concepts emerged which gave great importance to high tech in military affairs. NATO led in that direction, as witnessed by the adoption in the mid-eighties of the operational concept of Follow on Forces Attack (FOFA), promoted by then-SACEUR (Supreme Allied Commander, Europe), General Rodgers.

By the end of the 1970s, both the East and West came to realize that one had reached the threshold of a "military-technical revolution" that could dramatically transform warfare, calling for both new equipment as well as new defense organizations and doctrines. This convergence of opinion between the two opposed camps in appreciating the potential of new technologies for future warfare resulted however, from different approaches.

The Soviets tended to develop a more conceptual reflection of modern warfare that emphasized historical and conceptual considerations. Under the leadership of men like Marshall Ogarkov, General Gariev and General Vorobiev, the Russians modified their military doctrine in order to be able to swiftly defeat the Western camp in case of war on the Central Front, i.e. in Central Europe. They grounded their views on the need to conduct rapid decisive military operations into the full depth of enemy territory.[1] Soviet weaknesses in high technology led Soviet strategic thinkers to emphasize doctrinal innovation more than technological development. Developing "jointness" at the operational level of warfare, they sought to integrate their forces in the framework of "Reconnaissance Strike Complexes" and at the tactical level into a "Reconnaissance Fire Complex." They also innovated by creating new types of forces, such as the Operational Maneuver Group (OMG), aimed at launching raids into the depth of enemy territory.

The Americans responded to the challenge with more of an "engineering" approach, which led them to focus their analyses on the potential consequences of technological progresses for armaments development. At the same time, the U.S. armed services tried to develop innovative doctrines to cope with new Soviet approach to

---

[1] It is interesting to note that this notion of Rapid Decisive Operations (RDO) is the key organizing concept of the current U.S. transformation process.

conventional warfare, doctrine in which technological developments acquired an important role. During this period, Europe stayed away from that process. With the exception of France, which built its own nuclear force and defined a proper concept of deterrence—the "weak against the strong"—Europeans, by and large deserted the field of strategic and military thinking. Their military R&D was thus aimed at only seeking new traditional weapon systems with very few innovative programs. They left the quasi-monopoly of these affairs to Washington and Moscow even if at the occasion of the "Euro-missile crisis," some German thinkers did try to propose a reappraisal of the German military posture without success.

Currently, when Europe seeks to reevaluate its defense posture, particularly but not exclusively in the context of the ongoing U.S. defense "transformation", the lesson drawn from the Soviet experience should lead us to realize that strategic analysis and operational concepts are not necessarily only defined in accordance to mainly technological evolution. The assumption that technological evolution is the key factor is now being an assumption challenged by the difficulties the Western Coalition is confronted with in post Saddam's Iraq, where the U.S. high tech military is apparently unfit to efficiently cope with asymmetric threats. However, the Europeans would be foolish not to look carefully at what is going on in the United States in order to prepare future high intensity wars. Such undertaking is the goal of NATO's ACT, tasked to guarantee the conformity of Allied forces with those of the United States This task is also performed through a more confidential channel created in 1999, gathering few Western countries, within the framework of the Multinational Interoperability Council (MIC).[2]

Transformation: a Multi-Purpose Notion

Transformation has a different meaning in both the United States and in Europe. Accordingly, it has a different impact on military R&D on the two sides of the Atlantic. In America, the notion is derived

---

[2] The United States, Germany, the United Kingdom, France, Canada and Australia are members of the MIC, which was set up in 1999.aggression and mobility capabilities, and his protection. Félin will be operational in 2006. Its British equivalent is the FIST (Future Integrated Soldier Technology).

from the "Revolution in Military Affairs" (RMA), articulated in the early 1990's by Dr. Andrew Marshall and Admiral William Owens. Transformation has, however, more concrete applications than the RMA. The concept of network centric warfare, central to the idea of transformation, is a good example of how transformation may have a direct impact on the forces and the way they are interconnected. As such, transformation is betting on the tremendous capabilities offered by high tech systems to efficiently interconnect, almost in real time, sensors to shooters. Such a result is also a function of new chains of command that are grounded on vast amounts of assets and manpower dedicated to C4ISR. Transformation is also a notion which has been used to justify considerable increases in military spending, though the United States already spends about half of the world's total military spending and dwarfs any of its potential challengers—notably the countries which belong to the "Axis of Evil."[3]

In Europe transformation has a different meaning than in the United States. It is primarily seen as the adjustment of military posture to meet the totally new different security environment that has emerged after the Cold War. In most European countries, albeit at widely varying rates, forces are being made more mobile and tailored for expeditionary operations. As such, transformation is a radical departure from Cold War military structure, with less emphasis on high technology than in U.S. transformation plans. In some European countries such as France, forces have been "professionalized", reduced in number, but qualitatively improved. Command structures have been greatly transformed, most notably in the three most militarily potent European allies. It is worthwhile to point out that Berlin, London and Paris have committed substantial resources to set up command structures at both the strategic and operational levels, a capacity that no other NATO European countries possess. The three capitals have thus devoted a huge level of effort to create the PJHQ (Permanent Joint Headquarters—Northwood) in Britain, the CPCO (*Centre de Préparation et de Conduite des Opérations*—Paris) in France, and the EinsFüKdo (*EinsatzFührunsgsKommando*—Potsdam) in Berlin. These command structures are in direct communications with the MIC, where transformational concepts and projects are discussed

---

[3] The EU at 15 spends 150 billon Euros for defense. As a whole, it is the second largest military budget in the world.

among key allies. At the same time, these strategic headquarters are at the forefront of processes aimed at enhancing military planning and military capabilities to conduct EU military operations should the ESDP mature.

The State of ESDP

Nowadays, ESDP does possess an agenda. Its goal is to give the EU the capacities and the capabilities at the strategic level to independently assess a crisis, assess its potential military implications, plan military operations as necessary, and execute operations using purely European assets such as the newly-created ERRF (European Rapid Reaction Force). Relevant EU political-military structures are now able to conduct and supervise military operations. In addition, the EU finally reached an agreement with NATO regarding the implementation of EU-led military operations when using NATO assets (i.e., U.S. assets). This is the result of the recent (2002) completion of the Berlin Plus agreement. Those European structures have been tested in 2003 with two types of operations: one, *Concordia*, in the Balkans with the support of NATO assets; the other, *Artemis*, in the Democratic Republic of the Congo without relying on NATO assets; both were commanded by a French general.

In the field of military R&T development progress is slow. Resources are limited as indicated in the following chart:

Table 1. Military R&T spending of the 7 biggest contributors in the EU for 2002 (in M€)

UK	672
France	610
Germany	350
Sweden	133
The Netherlands	50
Spain	46
Italy	36
Total	1897

Besides organizational issues, it seems, however, that one can witness converging possibilities within Europe to pool R&D resources on future weapons systems, particularly in six domains:

- *Planning and conducting military operations.* For this purpose, new sensors, new means of collecting data and processing flows of information are increasingly needed. In that sense, it may be envisage developing cooperation between France and the UK on projects already in their R&D phase (SORA in France, ISTAR and DTC Fusion in the UK).

- *Network centric systems.* It may be worthwhile, in a European context, to promote R&D cooperation when various countries are already at the beginning of developing systems related to network centric warfare. For example, for ground forces, the French are working on requirements for digitalization of the battlefield through the *Bulle Opérationnelle Aéroterrestre* (BOA) program, the Swedes through LED SYNT, the Germans through IDZ and the UK through FRES. Indeed such programs can be devised in different types of R&D sub-programs for application to air and naval requirements: concepts and architecture; sub-systems such as vetronics (Vehicle Electronics), sensors, robots, directed energy weapons; new platforms; etc.

- *Programs already defined among Europeans in order to implement military operations within the framework of ESDP.* A force catalog has been defined to fulfill the so-called Petersburg tasks: the "Helsinki Headline Goal." Deficiencies have been highlighted that led the Europeans to establish a "European Capacity Action Plan" (ECAP). Ten projects groups have been created to harmonize R&D capabilities on those deficiencies in UAVs (France is leading the group), NBC defense (led by Italy), air refueling (Spain), mobile HQs (the UK), CSAR (Germany), etc.

- *Theater Ballistic Missile Defense.* The prime objective is development of an advanced alert system.

- *UCAV (Unmanned Combat Air Vehicle).* A European program has already been proposed by France, which has decided to commit 300 million euros for R&D.

- *Space.* Despite difficulties met in that sector, the recent signature of a memorandum on operational requirements for space based reconnaissance assets by five European chiefs-of-staff

(France, Germany, Italy, Spain, and Belgium) could promote further R&D cooperation. The memorandum calls for a two-step approach: regrouping assets and capabilities existing or already planned in each country (optical reconnaissance in France with Helios I and II, and the Pléiades satellites; radar observation in Germany with the SAR-Lupe satellites; and the Cosmos-Skymed satellites in Italy).

Network Centric Warfare and European R&D Efforts

The concept of "Network Centric Warfare" (NCW) is seldom used in France and in most European countries, even though from technological and technical points of view, work is progressing to give French armed forces the capacities and the capabilities to work and fight on the digitalized battlefield. However, the technical imperatives derived from the need to operate on a new complex battlefield have to be related to political and strategic considerations that impact on the overall understanding of the NCW concept.

In military affairs, there is a growing trend towards a faster tempo of operations. Rapid success on the battlefield destabilizes a slower adversary unable to cope with the momentum of the fight, thereby minimizing casualties and destruction, and meeting Western popular desires to limit the duration of large-scale military operations. France is part of the current transformation of most Western armed forces in order to give its forces greater latitude to operate on the twenty-first century battlefield. Each branch of the French armed forces is going through a more incremental rather than revolutionary development of NWC.

The basic idea behind the new way of warfare is based on collecting, processing and sharing pertinent information. As such, this imperative is part of the modernization of French armed forces currently being implemented under the five-year defense program from 2003 to 2008 (*Loi de programmation militaire—2003-2008, or "LPM"*). This law is part of a greater endeavor to build French forces in accordance with the 2015 model (*modèle d'armée 2015*). Three objectives have been assigned to the current LPM that are not immediately dedicated to NWC. The first objective is to restore the availability of equipment; the second, to speed up modernization and preparation of future equipment; and the third, to consolidate the "professionalization" of the armed forces.

Regarding the construction and the modernization of new equipment, the priorities of the LPM are the following:

- The deployment the third new-generation nuclear SLBM and the building of the 4th;
- The development of the new supersonic cruise missile (ASMP-A) with new nuclear warhead;
- The construction of a new high power laser for nuclear simulation;
- The acquisition of proper means of command, control and communications for giving France the capacity to be a framework nation in the context of EU-led military operations at the strategic (development of the CPCO); operational (*Syracuse III* military telecommunication satellite) and tactical levels (digitalization and modernization of telecommunication systems).
- Development and modernization of intelligence gathering systems: launching of *Helios II* reconnaissance satellites in 2004 and 2008; reconnaissance UAV (medium endurance); new ELINT/COMINT ship.
- Improvement of force projection with the commissioning of a second nuclear aircraft carrier, new transport aircraft (A400M transport aircraft), helicopter (NH90, Cougar) and two LHD (delivered in 2005 and 2006).
- Deep strike improvement with the delivery of 57 *Rafale* for the Air Force, 19 for the Navy, and 500 cruise missiles (SCALP-EG).
- The improvement of the land forces capabilities, in particular the delivery of 117 new *Leclerc* MBT, 37 *Tigre* attack helicopters, and 10 new *Cobra* counter-artillery radars. The new *Félin*[4] equipment for the infantry fight on the digitalized battlefield will be delivered (14,000 systems).

---

[4] The *Félin* program (communication and equipment integrated infantryman system) started in December 2000. The aim of the program is to optimize the abilities of a dismounted soldier by giving him a fully integrated and modular system intended to improve in a coherent and balanced manner his observation and communication capabilities, his aggression and mobility capabilities, and his protection. Félin will be operational in 2006. Its British equivalent is the FIST (Future Integrated Soldier Technology).

- Two *Horizon* anti-aircraft Frigates will be delivered, a third E2C-*Hawkeye*, and eight multi-mission frigates will be ordered.

- A renewed effort on military R&T development accomplished: 7,07 billion Eeuros will be spent during the LPM.

In this latter domain of R&T development the French will try to work on developing an "AirLand" system of NWC, already mentioned, the BOA. The BOA is aimed at developing a cooperative combat system for ground forces. It is built around different types of sensors based on different types of materials (mobile or fixed) capable of collecting and disseminating information. The "shooters" will used in all available information sources in a cooperative manner. In order to establish proper communications, a network will be created. A preliminary study n the feasibility of the IP Internet Protocol, or (IP) has been realized under the aegis of the *Délégation Générale pour l'Armement* (DGA). This study, known as ATTILA, realized by Thalès, was aimed at drawing up a master plan for migrating tactical and joint theater networks to Internet technology. The recommendation resulting from the Attila study has to be consolidated and embedded in new weapons systems. France, which is part of NATO's twelve nation TACOMS Post 2000 project, has submitted the result of the Attila study.

Conclusion

The United States is innovating by basing its mode of warfare on the intensive use of sophisticated and complex C4ISR systems. Should the Europeans follow the direction taken by the U.S. military in their current transformational process? Or, should Europeans invent their own "grammar of war" that might correspond more closely to their own visions of modern warfare based on historical experience and cultural assumption? This question is central to the so-called gap between U.S. and European forces notably in the field of military R&T development. The mirage of high tech is being sold to the Europeans as the only panacea to military problems, and it would seem that American views on future warfighting, as epitomized in the notion of network centric warfare, are automatically the standard views in Europe as well. Indeed, the Europeans have the military know-how and most of the technological expertise to develop high tech military systems by themselves. However they need to invent a model of warfare that is specifically tailored to the future needs of the European Union, and probably one that places less emphasis on the role of technology than the American model.

Europe has to work together on this to overcome the combined pressure of high costs and the extreme complexity of high tech weapons systems, particularly in terms of "enablers" such as complex C4ISR systems encompassing space assets, all which are increasingly out of the reach of any single EU country with defense budgets at their current levels. Joint development of these enablers through a better integrated R&D effort, will not only provide the tools facilitating the conduct of large scale-military operations in the future, but it will also shape the critical new "grammar of war" that Europe currently lacks.

Chapter 4

Leveraging Dual Use Technologies for Defense Needs

Kenneth Flamm

"Dual use" is a relatively recent buzzword applied to a very old concept. Essentially, it means technologies and products that have both military and commercial applications. A defense industrial or technology policy built around dual use subsumes taking technologies and products developed for military purposes and commercializing them (spin-off) or taking products and technologies developed commercially and applying them in military systems (spin-on).

Most recently, in the rarified air of the U.S. military acquisition system, the idea of producing both military and commercial products on the same production lines, or in the same production facilities (dual-produce), has also been added to the mix of dual use ideas used to discuss defense technology policy in the United States. This may be an artifact of the regulatory structure of the U.S. defense procurement system, however, since casual observation suggests that producing military and commercial products in the same facilities and lines is less unusual in foreign military industries than in the United States.

Early U.S. History

Dual use as a military industrial policy in the United States arguably goes back to the birth of the United States in the late eighteenth century. Under the influence of French military advisers, the U.S. Army sponsored the development of mass production techniques and mechanized tools intended to make possible the manufacture of firearms with interchangeable parts. Development of this manufacturing technology took place inside the Army arsenals in Springfield, Massachusetts, and Harpers Ferry, West Virginia, and in the shops and factories of private contractors hired under contract to work with the arsenals in developing this technology in the nineteenth century.[1]

---

[1] See David A. Hounshell, *From the American System to Mass Production, 1800-1932* (Baltimore: Johns Hopkins University Press, 1984), chapter 1; Merritt Roe Smith, "Army Ordnance and the 'American system' of manufacturing, 1815-1861," in Merritt Roe Smith, ed., *Military Enterprise and Technological Change* (Cambridge: MIT, 1985).

The technologies and skills honed by the arsenals and their industry contracts played a central role in the development of the U.S. machine tool industry and, in a classic example of "spin-off," directly fed the development of the "American system of manufactures" that attracted a steady stream of foreign observers from the middle of the nineteenth century onward.[2] These American manufacturing techniques pushed U.S. manufacturing to global preeminence, and created a trajectory that led from Colt firearms, to Singer sewing machines, to inexpensive bicycles and clocks, and ultimately, to that early marvel of twentieth century mass production, Henry Ford's Model-T assembly lines.

This was not the only area where military-sponsored technology was important to commercial industries in the nineteenth century. Army engineers played an important role in the administrative and standardization efforts that created the first pre-Civil War railroad network in the United States.[3] The U.S. Navy underwrote the development of high quality steel production technology in the latter part of the nineteenth century, in the United States. And, while little has been written about it, it is clear that "spin-on" played an important role for the U.S. military acquisition system, as inexpensive, high quality armaments and weapons systems were produced more cheaply using mass production technologies and manufacturing systems honed in high volume commercial markets.

World War I marked another important instance of "spin-off" in the United States. The U.S. aircraft industry—virtually nonexistent in this country despite the Wright brothers' early and well-publicized experiments—went from almost no serious mass production to numerous major producers of aircraft, entirely on the strength of wartime procurement contracts. The fortunes of the U.S. aerospace industry remained tied to government spending throughout the remainder of the twentieth century, with the federal share of aircraft, airframe, and aero-engine output surging above and below the 50 percent mark as government spending (including air mail contracts in the

---

[2] Hounshell, op. cit.; Nathan Rosenberg, "Technological Change in the Machine Tool Industry, 1840-1910," in Nathan Rosenberg, *Perspectives on Technology* (Cambridge: Cambridge University, 1976).

[3] See Charles F. O'Connell, Jr., "The Corps of Engineers and the Rise of Modern Management, 1827-1856,"; Merritt Roe Smith, "Introduction," in Smith, ed., *Military Enterprise and Technological Change* pp. 8-9.

first half of the century and NASA in the last, in addition to defense procurement) ebbed and flowed. Indeed, in contrast to other (later) U.S. "high tech" industry success stories, where defense saw its share of industry sales decline to smaller and smaller single-digit shares of total output, as commercial sales took off and drowned military requirements in an ocean of mass production, the military share of aerospace remains large and important relative to commercial sales to this day.

Post-World War II High-Tech Development

World War II arguably was the first global conflict whose outcome was determined by relative success in a race for accelerated development of leading edge technology, then applied to military systems. The Manhattan Project and the atomic bomb may be the poster child for wartime U.S. technology, but at least as important was the development of the modern electronic, digital computer by British and American scientists and its application to cracking encrypted German and Japanese signals traffic. Electronics technology was honed and refined and applied to a host of less dramatic but nonetheless highly important applications—radar, fire control systems, the proximity fuze, secure communications. Defense investment in technology (effectively, the entire preexisting U.S. R&D infrastructure was absorbed, greatly expanded, and applied to military problems during the war) soared, and technologies developed during the war became the foundations for many of the high tech U.S. industrial success stories of the latter half of the twentieth century. Aerospace, electronics, atomic power, information technology—these are some of the industrial descendents of that wartime U.S. investment in technology.

The semiconductor and computer industries are particularly good examples of the role of defense investment in nurturing and supporting the early phases of postwar high tech industry in the U.S.[4] Government in general, and defense in particular, paid for a large share of R&D and accounted for the bulk of the market for these products in the 1950s. In the 1960s, however, success in "spinning off"

---

[4] See Kenneth Flamm, *Creating the Computer*, (Washington: Brookings, 1988), chapter 3; *Mismanaged Trade? Strategic Policy in the Semiconductor Industry*, (Washington, Brookings, 1996), chapter 1.

these technologies into commercial products drove huge growth rates in commercial sales of computers and semiconductors. The defense share of sales dropped rapidly. As the commercial market expanded, and generated large investments intended to bring new generations of commercial products online, the government and military share of technology investment shrank. Government continued to play a major role at the leading/bleeding edge, however, where the latest and greatest technology could be translated (albeit at a relatively high cost) into a qualitative advantage in military and weapons systems. As technologies were refined and honed, of course, yesterday's leading edge became cheaper and ultimately found its way into today's commercial system. Though computers and semiconductors are today vast and huge commercial industries, defense sales and technology investment continue to play an important role in the market for the very largest and most capable computer systems (supercomputers) and the most exotic, high performance semiconductor devices. By the 1990s, however, government sales overall—albeit of continuing importance in high risk, leading edge areas—were a very small piece of sales by powerhouse U.S. high tech industries, like computers and semiconductors. The vast bulk of the market, and the technology investment that went with it, danced to the rhythm of commercial sales. Shrinking military procurement budgets in the 1990s further accentuated the diminishing role of defense acquisition in U.S. high tech. In addition, technological capabilities became globalized, with overseas producers able to match and in some cases exceed American high tech capacities.

Implications for Defense

These new and steadily more apparent realities had some significant implications for changing U.S. acquisition strategies in the 1990s. While it now seems obvious, perhaps the most important point was that technology and products that were widely available commercially worldwide could not be "differentiators" for U.S. military systems. There was simply no way to make access to a technology that was distributed globally, in high volumes, a unique source of military advantage for the United States.

One attempt to fight this trend was the creation of an increasing complex export control system. But the globalization of production systems and the spread of technical capabilities in a variety of high tech industries—particularly in computers and electronics—meant that export controls would work poorly. Even when competitive, high volume products were primarily produced by U.S. producers (like Intel, with microprocessors). The mere fact that many tens of millions of these components were being shipped to customers around the world meant that it was effectively impossible to deny an adversary access to this technology in any absolute sense. Export controls were doomed to work poorly, if at all, in these situations.

Another issue was that with increasing competition worldwide in commercial high tech, product cycles became shorter and shorter. Juxtaposed against this was a fairly rigid and inflexible military acquisition system that stretched development and production of new systems over decades. This raised the dangerous possibility that a nimble adversary might be able to introduce new technology derived from the commercial marketplace into a weapons system, not by having access to the technology first (since it was available more or less openly to all on world markets), but by *introducing* the technology into a working, deployed system first. This was a new element of risk.

Finally, over decades the U.S. acquisition system had developed into a process for generating requirements for new products and technologies, then finding producers able to translate those requirements into working systems. Cost was rarely made a major factor in systems design (though it would often work its way through the back, budget door into procurement volumes). The penchant for sticking with performance requirements, no matter what the cost, risked cutting military users off from the benefits of a much larger commercial market and the technology investments and scale economies that came with that vastly larger commercial market. Tradeoffs between cost and performance were finally to make their way into the formal decision loop in the 1990s, as the pressures of declining procurement budgets were felt.

Dual Use Policy Basics

Faced with these new realities, and under the strong leadership of Defense Secretary William Perry in the 1990s, the Department of Defense (DoD) made some significant changes in its procurement strategy. These changes were mainly to persist through later leadership changes.

First, where a differentiation vis-à-vis adversary was not possible, or not important, DoD was to buy commercial components and systems whenever it could. When a defense system was to operate in a harsh environment, commercial components and systems would be adapted to the extent possible through hardening or ruggedization. A corollary to reliance on commercial technology was that DoD needed to ensure that access to this technology was not going to be an issue for DoD. That meant having friendly (headquartered in the United States or friendly or allied countries) suppliers willing to give Defense early and assured access to the best available technology. It meant in some cases worrying about whether there was going to be a competitive market environment for commercial technology. DoD, like any other large customer, was concerned about being at the mercy of actions taken by a monopoly supplier of some closely held technology. If correctly and intelligently applied, use of commercial components and technology could be highly beneficial to military systems. It could lower cost, while possibly improving the pace of innovation in military products, as rapidly evolving commercial products were incorporated steadily into upgraded military systems.

Second, where differentiation was required, and where qualitative military advantage was essential, an alternative strategy was needed. This involved cultivating a responsive supplier base, hopefully including some of the same commercial firms pushing the leading edge in mass-market technology. The idea was to leverage off of commercial technology investments and production volumes, where the underlying technology was not defense unique, by making it attractive for those producers to address the military market. Acquisition reforms, and ultimately, a willingness to invest serious money in differentiation, were key ingredients in this effort. The hope was, again, that an intelligent policy would leverage commercial technology strengths into systems differentiators by making it interesting for commercial firms to address defense markets. This might involve cases where there ultimately might be long term benefits to commercial sales from leading edge technologies brought in early to defense systems, and it might involve cases where Defense was simply willing to pay whatever it took to bring some capability to the DoD market early. The policy also recognized that there were defense-specific technologies and systems that were simply not relevant to commercial markets. These capabilities would simply have to be maintained. The policy issue in these cases boiled down to how to balance economies of scale against the benefits of competition, when considering how to maintain that costly, defense-unique industrial base.

Plan B

Dual use was not the only conceivable strategy to address the changing technology requirements and environment faced by DoD. In many respects, the simplest course of action was to fund the development of a parallel (where there was a counterpart commercial industry), leading edge defense technology base among the military contractors Defense had historically used for its systems. Needless to say, military contractors often saw little fault in this alternative. If one was willing to pay the price, went the logic, there was little or nothing that could not be invented or reinvented within the safe and secure boundaries of a defense contractor.

There were some disadvantages to this plan, however. First, costs might very well be substantially greater, given the limited size of military relative to commercial markets, where there were connections to commercial products that could share some of the R&D costs of developing new technology. Second, an isolated military contractor base, even if spending large amounts, risked being left behind by a much larger and fast-moving commercial mainstream in highly dynamic technology markets. If one could somehow lure commercial high tech firms into setting up military production divisions, this risk might be lessened. But it was hard to see, even with vast expenditures by DoD, how it was going to be in a mainstream high tech firm's economic interest to divert its best design talent and production resources into a military market that was only a single digit percentage of a vastly bigger commercial market.

Despite these disadvantages, however, a defense-unique industrial base may be the only alternative when military systems and products are substantially different from anything sold by commercial industry. In this case, we have classic issues of defense industrial policy that DoD has wrestled with for years. Most of these boil down to the economic tradeoffs between economies of scale (which argue for fewer firms with lower average costs paid by DoD) and competition (which argues for more firms in a quest for lower profit margins tacked onto costs and greater pressure for technological competition and resulting innovation). Submarines, fighter aircraft, aircraft carriers, missiles, electronic warfare, etc., are classic examples where there really is no analogous commercial market. Dual use, in these cases, at most would focus on encouraging the use of commercial components by those with defense-unique, system-integrating capabilities.

Dual use Reborn

These ideas were in many respects a reinvention of earlier experiences. In the 1960s, for example, J.C.R. Licklider, at DARPA, explicitly pioneered many of these concepts in the context of support for developing advanced computer technology. Licklider explicitly sought to develop high-risk technology in partnership with commercial firms (though we must note that a much larger share of commercial computer firms' sales went to Defense in that earlier era), so that DoD would benefit from early access to these products when they went into production. Internal DARPA reports of the mid-1960s argued that acceleration of technical progress within commercial companies would improve the quality of commercial products, and benefit DoD as a major customer for these leading edge systems.[5] A corollary to the argument was that because commercial technology was relatively cheaper than defense-unique designs with roughly similar capabilities, Defense would be able to make much more widespread use of the new technology in military systems. To slightly restate the argument in 1990s terms, "early access to the technology was what potentially made this sort of Dual use technology a differentiator for the U.S. military. Given long military procurement cycles, then and now, such early access was critical to making dual use technology investments a viable strategy for DoD."

"Plan B" (a defense-unique internal clone of a commercial high tech industry) had also been tried in earlier years. The classic example was DoD's Very High Speed Integrated Circuit (VHSIC) program of the 1980s, in which DoD invested almost $1 billion to accelerate innovation in the semiconductors used within military systems, which gained an advantage for military systems making use of these parts. The late 1970s were also years of radical change in a global semicon-

---

[5] Kenneth Flamm, *Targeting the Computer*, (Washington: Brookings, 1987), pp. 51-75.

ductor industry previously dominated by U.S. producers. Japan launched a series of government-industry semiconductor R&D consortia—the so-called Very Large Scale Integrated circuit (VLSI) projects. These efforts were perceived by most observers to have greatly advanced the technological and manufacturing competence of Japanese semiconductor producers, and VHSIC was perceived to be at least in part a response to these changes. Confined largely to defense contractors and the defense contracting divisions of commercial firms (who were uninterested in the program, given their own internal efforts in the same areas and the onerous security restrictions on information flows coming out of the VHSIC program), the VHSIC effort created very little tangible benefit for DoD or its defense industrial participants. Indeed, technically more advanced commercial products leapfrogged VHSIC products that were ultimately produced. (I note that there were a few useful things produced by VHSIC—some advanced semiconductor manufacturing equipment, for example, and the VHSIC Hardware Description Language, VHDL, but these benefits were tiny in relation to the enormous cost of the program and its minimal effect in pushing DoD microelectronics ahead of commercial products.)

In stark contrast to VHSIC was a later, explicitly dual use DoD initiative, SEMATECH. In 1987, the U.S. Defense Department's Defense Science Board issued a report noting a rapid deterioration in the relative position of American semiconductor manufacturers and characterizing this as a national security issue. In response, the U.S. government decided to have the Defense Department pay half of the cost of a joint industry consortium—dubbed SEMATECH (for semiconductor manufacturing technology)—and budgeted $200 million annually. In stark contrast to VHSIC, after 10 years and another billion dollars in DoD investment, U.S. semiconductor manufacturers have once again reclaimed a position at the leading edge of semiconductor manufacturing technology. Indeed, U.S. semiconductor capabilities have largely ceased to be an issue in national security policy debates. While the U.S. semiconductor renaissance is not just the result of SEMATECH's efforts, there is widespread agreement that SEMATECH deserves some reasonable share of the credit. SEMATECH even provided a forum in which the global (not just U.S. companies) semiconductor industry has developed a unique framework for coordinating international technology investments (the International

Semiconductor Technology Roadmap), a development associated with a visible speedup in the rate of technological innovation in this important industry.[6]

While the United States retains access to leading edge semiconductor manufacturing capabilities, these same capabilities are also available to others in this now globalized industry. Advanced semiconductors are not a unique differentiator for U.S. military systems as they were, say, in the 1960s. Nonetheless, other pieces of an increasingly international business do provide US systems producers with potential differentiators, particularly advanced US capabilities in systems integration and chip design. The moral is that the U.S. military has successfully adapted to changing industrial circumstances, and loss of what may once have been a differentiator has not created a crisis.

Defense Technology Cooperation with Europe: The Role of Dual Use

"Dual use" may occupy a particularly important role in enabling future cooperation in defense systems acquisition with our European partners and allies. To begin, our European allies, like most industrialized nation states, seek to maintain some autonomous capacity to design and build the means of defending themselves against external aggressors. This means retaining the corresponding defense industrial base needed to supply a certain degree of credibility should they seek to play a role in the resolution of global or regional security problems in which they have a perceived national interest. At the same time, the United States spends vastly greater sums on the development of new defense technology, and this creates a fundamental asymmetry in the security relationship between the United States and Europe. **Figure 1** illustrates the magnitude of this divergence.

---

[6] See Kenneth Flamm, "The New Economy in Historical Perspective: Evolution of Digital Electronics Technology," in Derek C. Jones, ed., *New Economy Handbook* (New York: Academic Press, 2003).

Figure 1. Defense R&D, 1999

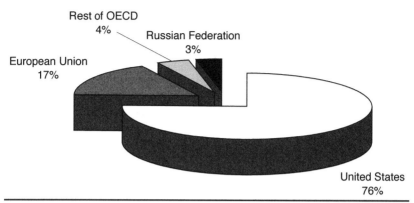

Source: OECD database.

The United States spent almost $4.50 on defense R&D for every dollar spent in the European Union in 1999. Roughly the same pattern could be observed in 2000 (see **Figure 2** below), with $4.30 spent by the United States for every $1 in defense R&D spending in the European Union. This is a very large differential—so much so that its sheer magnitude creates obstacles to Transatlantic cooperation on technology projects.

Figure 2. Defense R&D, 2000

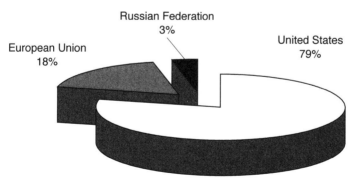

Source: OECD database.

In principle, both sides of the Atlantic gain by fielding common equipment in their armed forces, which potentially makes joint operations substantially easier to support. The desire for Europe to maintain autonomous, viable, defense capabilities, however, means that

Europe is unlikely to simply field systems in US defense industry markets, whether or not these systems are manufactured in some form of partnership with local European producers. Through U.S. eyes, fielding less capable equipment—on average, at least, an inevitable consequence of investing substantially less in its development—is not an acceptable alternative. To European eyes, becoming wholly reliant on defense technology imported from abroad in the form of imported US equipment designs, even if manufactured locally in partnership with an American defense industrial partner, is equally unacceptable. How then can a common interest in fielding common equipment be served, yet bridge these divergent interests?

Historically, cooperation on specific joint, Transatlantic defense systems projects has been attempted in order to bridge this divergence in objectives. European partners would contribute their expertise in specific areas where they had chosen to specialize in defense technology. By bringing rough parity in capability to bear on these select projects, despite the overall imbalance in defense technology investment, the foundations of a more equal cooperative relationship would be laid. A bona fide contribution of technology, rather than a relatively small investment of procurement budget, would make for a more equal relationship and make the U.S. defense R&D giant more willing to share technology. Instead of tolerating European technology partners in the interests of Alliance cohesion, by halfheartedly participating in joint projects of marginal interest to the uniformed services (while undertaking truly important projects alone), the United States would be motivated to work on genuinely significant joint projects.

Access to unique European ideas and expertise, not already available at home within its domestic defense industrial base, would be DoD's incentive to work on truly significant joint development projects in a partnership of approximate equals. Under these circumstances, it was hoped, the United States would be less reluctant to transfer serious chunks of its own technical know-how, so critical to European aspirations to a leading edge defense industrial base.

This was the theory underlying dreams of transatlantic defense industrial development projects. Arrayed against this theory, however, is the tyranny of the R&D spending numbers. While European defense industry certainly has some great "nuggets" of defense technology (a good and original idea can be as or more important than vast sums expended in a less effective or creative way), the sheer magnitude of the differences in size for investments in defense technologies between the United States and Europe makes the search for areas of rough parity in defense technology challenging. A more promising approach might be to bring more substantial or roughly equal European investments—and strengths—in dual use technologies to bear on the problem of stimulating genuine R&D partnerships in defense. "True" R&D cooperation is simply going to be easier in areas where both sides face a rough symmetry in the potential technical gains from cooperation.

If transatlantic R&D cooperation is to mean more than carefully controlled release of technology crumbs from projects building on massive U.S. investments in defense technology, a release calibrated to provide the minimum technological incentive required to seal the deal on licensing or sale of U.S.-designed technology and systems, some strategy that levels the technological playing field is needed.

What are the alternatives that might give Europe some genuine leverage in seeking a substantial place at the table? One argument might be that European equipment procurement is substantial enough to provide the needed balance and that the European defense market is a big enough bargaining chip to induce the United States to adopt a more accommodating stance toward European desires for technology sharing. Certainly the disparity between the United States and Europe on the scales of procurement investment is much less pronounced than the defense technology imbalance. **Figure 3** shows that the United States procures about $2 in equipment for every dollar of European equipment, a substantially more equal comparison than was the case in defense R&D.

Ironically, this imbalance in equipment procurement, though less than in defense technology investment, aggravates rather than improves Europe's problem. Given large economies of scale in many defense industries as well as economies of scale linked to large, relatively fixed R&D requirements for modern, leading edge systems, this differential in procurement volumes spells a large unit cost disadvantage for European defense producers compared with their American competitors. Even if Europe managed to unify its still overly fragmented defense industries into a handful of "Fortress Europe" mega-firms, the same unit cost disadvantage would remain. This places greater pressure on European defense producers to export their systems in order to drive up volumes, thus making it even more essential that European firms field systems with technology comparable to those shipped by American competitors.

Figure 3. NATO Equipment $, 2002

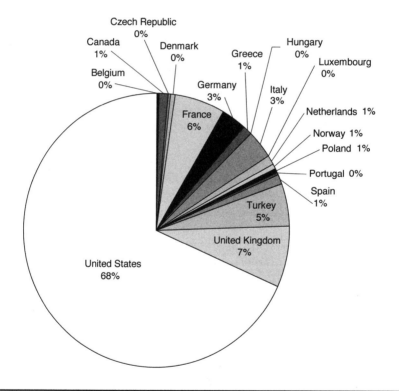

Source: calculated from NATO statistics. Converted to $ using OECD PPPs.

A much more promising direction in which to seek rough parity in bargaining chips is in civil space programs. Both Europe and Japan are investing serious and growing sums in their indigenous space capabilities. **Figure 4** (below) shows that in 1999, civil space spending amounted to less than half of the U.S. total.

Figure 4. Civil Space R&D, 1999

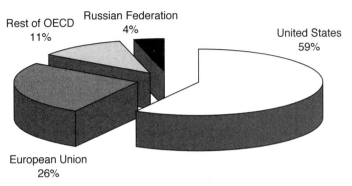

Source: OECD database.

European investments in space were already accelerating rapidly at the turn of the century however. By 2000, European civil space R&D had grown to three-fourths of the U.S. investment. Today, we are approaching a situation of rough parity.[7] (See **Figure 5**.)

Figure 5. Civil Space R&D, 2000

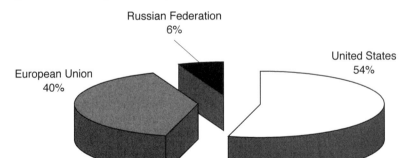

Source: OECD Database

---

[7] Some European analysts argue that European civil space R&D figures used here misrepresent the dimensions of the European effort, since French statistics, for example, seem to categorize many "one-off," prototype development efforts as development, while U.S. programs producing larger numbers of a single design do not count systems past the initial prototype in R&D. Because the U.S. produces a relatively larger number of production systems, goes the argument, a larger fraction of the total spending on space gets counted as R&D in Europe. Therefore, it is argued, the apparent European catch-up in civil space R&D is to some extent an optical illusion, driven by the fact that most European systems are prototypes, in contrast to the U.S., where substantial numbers of systems are built in production volumes.

There is a thin line between civil and defense in many space applications. Communications, imaging and surveillance, and navigation aids are areas where civil requirements and military needs can be reasonably similar. In other areas, of course, the line between civilian and military requirements is much brighter. Signals intelligence and space platforms for military systems, of course, are areas where military requirements are primarily defense-unique. Still, the potential relevance to military applications of civil space spending is great. If we add civil space R&D to defense R&D, in 1999, the United States still spent almost $4 for every European R&D dollar. (See **Figure 6**.)

Figure 6. Defense + Space R&D, 1999

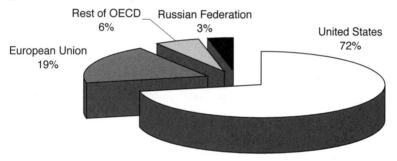

Source: OECD Database.

But with the increasing pace of European spending on space, this differential narrowed rapidly. By 2000, the U.S./European differential in combined defense and space spending had dropped to more like a 3:1 ratio. (See **Figure 7**.)

Figure 7. Defense + Space R&D, 2000

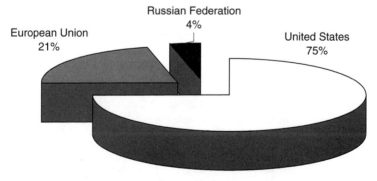

Source: OECD database.

Joint systems projects in the space arena—particularly in communications, imaging, and navigation—would seem like useful areas to explore in the quest for candidate systems where both sides spending and capabilities would be more evenly matched.

Europe has important strengths in other dual use areas important to defense. Information technology is one such area. European producers are global leaders, with world-class technology, in communications hardware (**Figures 8** and **9**), and in embedded software (**Figure 10**). In civil aerospace and aero engines, European firms are true peers with their American competitors. In biotechnology and pharmaceuticals, increasingly important to defense needs in an era of threats from terrorists and asymmetric opponents, firms in Europe are leaders in their industries.

Figure 8. Transmission Equipment Production, 1995–1998

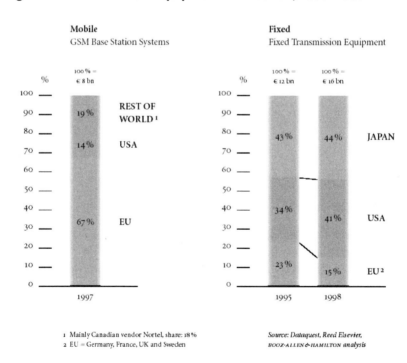

Source: European Commission, European IT Observatory, 2002.

Figure 9. Switching Equipment Production, 1995–1998

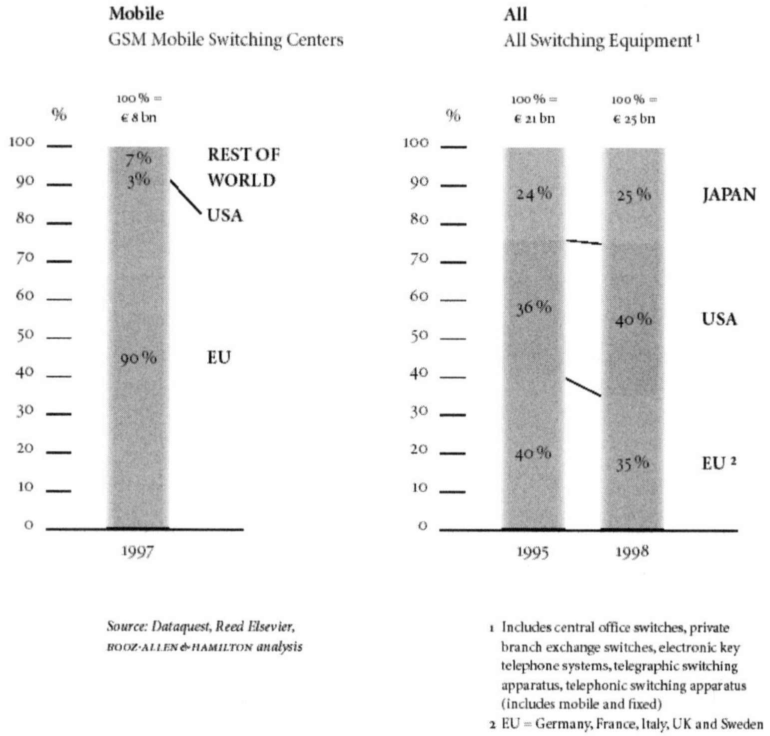

Mobile
GSM Mobile Switching Centers

1997 (100% = €8bn):
- REST OF WORLD: 7%
- USA: 3%
- EU: 90%

All
All Switching Equipment [1]

1995 (100% = €21bn):
- JAPAN: 24%
- USA: 36%
- EU [2]: 40%

1998 (100% = €25bn):
- JAPAN: 25%
- USA: 40%
- EU [2]: 35%

Source: *Dataquest, Reed Elsevier, BOOZ-ALLEN & HAMILTON analysis*

[1] Includes central office switches, private branch exchange switches, electronic key telephone systems, telegraphic switching apparatus, telephonic switching apparatus (includes mobile and fixed)
[2] EU = Germany, France, Italy, UK and Sweden

Source: European Commission, European IT Observatory, 2002.

Conclusion

Dual use means tapping into commercial technology solutions where appropriate. It has long been a feature (albeit an unadvertised one until fairly recently) of American defense industrial policy. It is not feasible in all areas nor does it solve all problems. But it does potentially provide the basis for more symmetric contributions to technology cooperation projects of interest to both European and American defense establishments. More equal contributions would most likely spell a more equal distribution of technical benefit, which is of vital interest in a Europe determined to maintain its defense industrial capabilities at the leading edge.

Figure 10. Total Embedded Software Production, 1995–1998[1]
Estimates

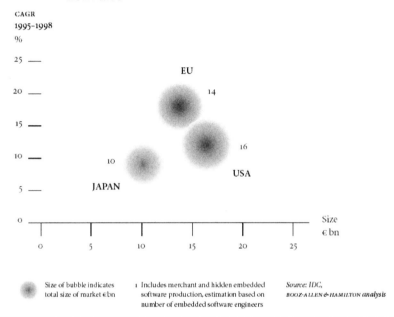

Source: European Commission, European IT Observatory, 2002.

Dual use areas with potential for productive collaboration on defense-relevant systems would seem to include space, aero structures and engines, software, communications, and biotechnology and pharmaceuticals. Not surprisingly, these are also areas where there is already a substantial web of commercial collaborations between American and European producers. Expanding this network to defense projects would seem like the path of least resistance to follow in fostering a more attractive mix of technology development projects, heightening their importance to both sides, and fostering deeper defense cooperation.

Chapter 5

The EU's Role in Leveraging European Multiple Use Technologies for Defense Needs

Klaus Becher

Since the creation of its first institutional components more than half a century ago, the European Union has been a laboratory for many different forms of cooperation between nations in their effort to improve their respective ability to cope with problems more effectively. In some aspects, such as external trade policy or monetary union, this has led to complete integration, i.e., the transfer of traditionally national functions and legal authority to the jointly established political and administrative community bodies. In research, however, such an integrated approach has not been established. This is ironic given the key role of research and development as both a driver and potential destabilizing element in the EU's development of a Single Market economy, the unhindered mobility of research-related capital and labor, and the EU's position in the world.

The European Commission has come to play a considerable role in public research funding, notably through its five-year framework programs. At the same time, however, the research role of individual members states has not diminished. The EU's draft constitutional treaty of 2003 contains clauses that would ratify and perpetuate this situation, a legally unique one in the overall EU framework. For research, technological development and space, both the Union and the member states are going to be in charge, and even if the Union exercises its shared jurisdiction, member states can still continue to use their own as they wish.

It should be recognized that European states have during the last half century been pursuing a good portion of their national (non-EU), publicly-funded research and technology development activities in various international cooperation schemes and organizations, such as the European Space Agency (ESA), which has a different membership than the EU and, at least so far, independence from the EU. There is also a growing trend to bring non-EU elements of the overall European integration process organizationally closer to the EU. For example, the long-standing intergovernmental framework Cooperation in the field of scientific and technical research (COST) is now served by the EU Council's General Secretariat, although its 32 member countries include many non-EU members. Today, 80% of public research spending in the EU is national, and merely 5% is done through the EU. The preference for continuing multiple approaches to research policy may partly be caused by the strong survival instincts of existing national research establishments, but more importantly, it reflects a predominant attitude that research should not be centralized and should not be driven primarily by bureaucrats.

The downside has been a noticeably disjointed approach to research in the EU as a whole, with a multitude of public agencies pursuing overlapping and fragmented agendas of their own instead of taking best advantage of the research funds expended. The large amount of intra-EU duplication and the absence of common objectives have been more and more deplored as damaging to the EU and its members in recent years. At this juncture, the EU Commission sees this as its role to act as a catalyst for change and to take initiatives geared at transforming the European research landscape.

European research and technology development efforts over the last three decades have often been a success story, at times creating major new commercial opportunities. The list of well-known European achievements ranges from particle physics (and the creation of the world wide web) at the European Council for Nuclear Research (CERN), the Ariane family of space launch rockets, and the Airbus line of large passenger and cargo aircraft, to lesser known strengths such as advanced radar and mobile communications components, building on a wide, well-maintained culture of high-quality engineering and systems integration. Many of these successful efforts were motivated, to a considerable degree, by the need to catch up with U.S. technology, and where possible surpass it, in areas where the U.S. government and industry exploited technological advantages unfairly, e.g., by refusing to launch European commercial satellites on U.S. launchers in the 1960s and 1970s.

In recent years, the EU governments and the European Commission have set ambitious goals for expanding the pattern of past successes to underpin the economic vitality and political weight of the expanded European Union, soon to include 450 million people in 25 countries, representing more than one quarter of the world's GDP. In March 2000, the European Council defined the political objective to "make Europe the most competitive and dynamic knowledge-based economy by 2010" in an effort to launch a new era of job-creating growth in Europe's lagging economies and strengthen international competitiveness through focused investment in the knowledge economy. At the European Council meeting in Barcelona in March 2002, the EU committed itself to an effort for strengthening industrial performance and "closing the gap between the EU and its major competitors" by raising total R&D spending in the EU economies from 1.9 to 3% of GDP by 2010, with two thirds of the total hoped to come from industry. Particular emphasis was to be put on "frontier technologies."

On this basis, the European Commission presented an action plan in April 2003 that suggested better coordination of national efforts, for example, through the creation of "European technology platforms" that assemble the main stakeholders (research organizations, industry, regulators, users, etc.) around key technologies, and increased public research expenditure geared at creating "abundant and excellent teams of researchers."[1]

The EU's Involvement in Security

During the Cold War, good reasons existed for keeping defense-related research separate from civilian research. This applied at the national level, with relatively minor differences between individual countries, and also on the level of European integration. The oldest shared European research center, CERN, was designed to "have no concern with work for military requirements" when it was established in the early 1950s.

The ESA, in Article II of its Convention, is limited to "exclusively peaceful purposes"—a phrase that has until very recently been interpreted as excluding any defense-related programs (contrary to the

---

[1] Commission of the European Communities, *Investing in Research: an action plan for Europe, Communication from the Commission*, COM(2003)226 final, Brussels, 30 April 2003.

meaning of "peaceful purposes" as "non-aggressive" in the Treaty on Outer Space, but consistent with the meaning of the same phrase as "non-military" in the nuclear proliferation context). This restriction, however, never limited the right of member states to use ESA's facilities and launchers for any kind of national programs.[2]

After a debate originally triggered in 1990 by the discussion in the Western European Union (WEU) about a future European observation satellite system for verification, ESA's traditional policy of self-restraint by means of narrow legal interpretation of its constitution is now finally being challenged from within. The goal is to enable the agency to offer its services for a coherent, "multiple use" space policy, thus helping to avoid the need for establishing a separate European military space agency that would duplicate existing capabilities.

The ESA is also attempting to put the necessary security regulations and facilities in place for handling classified projects. In the past, its technical facilities at ESTEC in the Netherlands have already been used for testing Helios-1 components. Developments such as the hyper-spectral imaging instrument built for Envisat clearly have multiple use applications, as has optical communications link technology, demonstrated successfully with ESA's advanced communications technology platform Artemis.

In the European Community's treaty language, no ban on security-related activities exists, although budget rules and political sensitivities create a number of obstacles. One further limitation is that the European Community Treaty (Art. 296) preserves the right of members states to exclude defense production and trade (but not dual use goods) from the Community's single-market and competition policy on grounds of essential national security interests[3]—a rule that is today seen as increasingly problematical, given the cross-spectrum activities of many defense-industrial companies, integrating civilian and defense applications, and also the increasing transnational nature

---

[2] Cf. Wulf von Kries, "The ESA Convention and Europe's Security Interests in Outer Space," *Zeitschrift für Luft- und Weltraumrecht*, 2/1994.

[3] "Any Member State may take such measures as it considers necessary for the protection of the essential interests of its security which are connected with the production of or trade in arms, munitions and war material; such measures shall not adversely affect the conditions of competition in the common market regarding products which are not intended for specifically military purposes."

of the large private-sector players in this market. On the political level, an additional limit on EU ventures in security-related fields results from the long-held ambition to shape the EU into a "civilian power" that somehow keeps its distance from all things military. This notion was in the past fed by various factors, but had its main intellectual root in the hope, clearly obsolete since 1991, that Europe could steer its own way between the two antagonistic nuclear superpowers during the East-West conflict.

EC officials and institutions have cultivated an explicitly civilian identity, seen by many as a progressive alternative to allegedly "old-fashioned" political notions like defense. It would be an illusion to believe that this attitude has really changed all that much yet, also not in the research field. There is both a continuing reluctance of EU institutions to get involved in anything but the softest aspects of security and a continuing preference on the part of member states to keep this realm under national control as long and as much as possible. For the longest time member states have been very reluctant to recognize any security role of the European Commission. As far as defense is concerned, this is still true today and is likely to remain true for the foreseeable future. NATO's long preponderance in security policy and defense and the desire to preserve national sovereignty over security and defense left little room for an EU role.

In a different way, though, security is really nothing new for the European Union. One of the original Communities that have now been merged into the EU—EURATOM—while not dealing with military nuclear programs, was always concerned with security issues such as proliferation safeguards and nuclear waste management. And of course in a wider sense, the institutions of European integration as a whole, created in the 1950s, resulted from the desire to provide security for Europe, both against the communist threat and the resurgence of war between European nations.

Remarkable change towards a stronger security focus occurred in the dynamic process of increasing cooperation in foreign policy among member states and the gradual emergence of the EU as a major international actor since the 1970s. In the mid-1980s, economic aspects of security became legitimate topics for European foreign policy coordination. In the 1990s, the Balkans wars brought home the lesson that money and good will alone were insufficient to prevent and end wars and protect Europeans from violence and its consequences. The necessary role of military "tools" has been more widely accepted since then, allowing a more open-minded discussion on security and defense within the EU.

While these rapid developments mainly took place in the European Union's second pillar Common Foreign and Security Policy (CFSP) and also third pillar (justice and home affairs), the European Commission is now also to some degree seizing the opportunity to address security issues in its external relations and humanitarian aid activities in a bid to complement the national policies of member states and their common efforts under the European Council. Only very recently has this trend been extended to Commission activities such as research, information technology and space. In 2001, the European Commission established a European Advisory Group on Aerospace. Its July 2002 report "STAR 21: Strategic Aerospace Review for the 21st Century—Creating a Coherent Market and Policy Framework for a Vital European Industry" marked the beginning of a new era of security and defense-related policy-making in the European Commission by relating the development of defense-industrial markets with the new challenges that result from the political decision to strengthen the European Security and Defense Policy (ESDP).

Another key event in this context has been the decision of member states, in 1999, to urge the European Commission and ESA to work together to develop a coherent "European space policy". After a productive series of joint efforts for the conceptual development of such a coherent approach to space, this rapprochement between the EU and ESA has led to the conclusion of a Framework Agreement for structured cooperation between the two bodies in November 2003. The agreement recognizes the specific complementary and mutually reinforcing strengths of both institutions and commits them to working together while avoiding unnecessary duplication. The EU and ESA have committed themselves to the development of a joint R&D work plan and an appropriate management structure.

Specifically, under the framework agreement, it will be ESA's role to supply space systems and infrastructures necessary to meet the demand for space-related services and applications for EU policies. With a view to the security dimension of these EU policies, especially ESDP, it is hoped that this arrangement with ESA will eliminate the need to establish a wasteful separate European agency for military space cooperation. Also in November 2003, the EU presented its White Paper on space,[4] an action plan for the coming decade drafted on the basis of a "Green Paper" consultation process that reached out to the dispersed user communities of space systems and applications in the EU.

New Approaches to Security-Related Research

There is now broad consensus in Europe on the pragmatic argument that the civilian image of EU activities is worth preserving. This is also the case for the security aspect of the EU's research activities. Required research funds may indeed in many cases be easier to mobilize in the name of civilian and dual use objectives in the European political arena, both nationally and on the EU level, than under an explicit security and defense heading. The shrinking defense budgets of the 1990s, and especially the severe squeeze on defense R&D spending at a time of rapid technological change in defense and an increasingly ambitious agenda of European security and defense policy, make it almost inevitable that governments look for possible ways to tap into non-defense budgets, both on the national and the EU levels, for indirect support of security and defense activities.

In Europe, public investment in R&D is only a fraction of what the U.S. government is spending in this field, with the majority coming from the DoD budget. While the UK and, to a lesser degree, France are almost in the same league with the United States in terms of per-soldier defense spending, defense R&D, and procurement, Europe as a whole suffers from a lack of defense-related research investment. This has obviously put European industrial firms in defense and certain high-technology areas at a structural disadvantage *vis-à-vis* U.S. competitors, particularly in fields such as aerospace where major industrial players are active in defense as well as non-defense applications.

In defense, Europe's research effort has been even more pointedly national than in research as a whole. Very few bi- and multi-national approaches exist. The most important one so far is the Western

---

[4] Commission of the European Communities, *Space: a new European frontier for an expanding Union. An action plan for implementing the European Space Policy, White Paper*, COM(2003)673, Brussels, 11 November 2003.

European Armaments Group (WEAG), which grew from the earlier Independent European Programme Group (IEPG). It provides a flexible framework for shared research and technology development efforts on the basis of à-la-carte cooperation within a number of standing thematic working groups. Compared to national defense research, only a very small fraction of ongoing efforts is involved, and the financial dimension is marginal. WEAG formed part of the WEU but, like the WEU's parliamentary assembly, was not transferred to the EU when the two organizations were merged in 2001. At the time, the wide consensus that exists today among EU nations about the need to boost European defense-research cooperation had not yet emerged.

A European Armaments Agency and other New Tools. In the meantime, a considerable degree of policy convergence on this issue between the major EU member states has led to the decision by the European Council in Thessaloniki in 2003 to work towards the establishment of a European Armaments, Research and Military Capabilities Agency as a new component of the EU's security and defense policy mechanism. The EU's draft constitution includes provisions for such an agency. On the part of the EU, the effort to foster better research support for European security and defense policy is embedded in a much wider political impulse that seeks both a strengthening of European research and its better direction at the support of practical European policy priorities. The EU's research policy as a whole is being refocused from a technology-driven approach towards one driven by applications and missions. In EU thinking, research is now meant not just to strengthen competitiveness and employment but also providing operational policy support. While in the past this would have mainly meant the EU's agricultural or environmental policy, it now increasingly also includes external aid, humanitarian relief, emergency response, conflict prevention, crisis management and "peace operations" of various kinds.

In this context, the Commission's Joint Research Center (JRC) has been reshaped. One of the innovations was the establishment of an Institute for the Protection and Security of the Citizen (IPSC) at JRC with 400 staff and a 60 million euros annual budget. Today, a broad range of inherently security-related research fields are being pursued at the JRC. This includes not just nuclear nonproliferation and safeguards but also such fields as humanitarian de-mining, coastal monitoring, information security, biotechnology and implications for terrorism, intelligence technology for strengthening anti-fraud measures, and open-source intelligence software. The EU's 6th Framework Program for 2002-2006, managed by the Commission, for the first time attempts to reflect the new significance of security aspects. Aeronautics and space research have been designated a thematic priority, with GALILEO, GMES and satellite communications as the focus areas. More than 1bn euros are earmarked for EU-supported aerospace research over 5 years. This is still less than what is spent on space research and technology development in ESA over the same time. However, it represents a significant increase in overall public spending for applications-oriented research in the EU and its member states in this sector.

The important issue here is that the Framework Program is not an intergovernmental arrangement for pooling and distributing national research money on the basis of "fair return". Instead, it offers additional funds from the EU's own budget, which flow from the Community's own revenues. In a situation of extremely tight restrictions on national discretionary spending, this may give substantial appeal to the notion of using the EU to mobilize bitterly needed funding for dual use purposes that also benefit security and defense. While the term "defense" is still generally avoided in this context, with the exception of occasional references to ESDP and the "Petersberg tasks" (now incorporated in the EU Treaty), the new reality of EU-led military operations, such as in 2003 the small-scale missions in Macedonia and Congo, offers a legitimate basis for the EU as a whole to work towards defense preparedness.

The practical need to provide required tools for ESDP gives focus and direction to the emerging security dimensions of EU research policy. The fact that ESDP encompasses both the military dimension (for a wide range of contingencies from humanitarian assistance to combat operations) and a civilian dimension (such as the current police operation in Bosnia-Herzegovina) serves as an additional element of justification for the EU's preferred multiple use approach to research and technology development for security and defense. This is further reinforced by the fact that in today's strategic environment, external and internal security are increasingly seen as part of one wider concept. The same technologies, such as sensors or secure communications, are helpful for both dimensions.

The EU's security-related research agenda is expected to better enable member states to act on the shortfalls identified and commitments made under the European Capabilities Action Plan (ECAP), which focuses on a so-called "capabilities gap" and how Europe should address it. The new research agenda can help member states to consolidate their national plans, redirect investment plans to bring them in line with emerging, shared requirements wherever possible, and avoid wasteful intra-European duplication, such as those currently pursued, for example, in the field of space-based radar imaging with the German SAR-Lupe and the Italian Cosmo-Skymed projects.

Better use of civil-military synergies and "multiple use" possibilities can provide an important key to significant capabilities improvements at a time of scarce financial means and uncertain mission scenarios. The satellite center in Torrejón, which the EU inherited from the WEU, though not equipped and organised to play a fully operational role in ESDP missions, demonstrates this quite well by exploiting commercial and other imagery for a wide range of security and defense purposes.

As national research efforts in most countries have traditionally suffered from a strict separation between the civil and military domains, the EU's "multiple use" approach is likely to also have some beneficial influence on national debates and practices. At the request of the European Parliament, the Commission recently presented its vision for an EU defense equipment policy,[5] pointing to the significant economic efficiencies that an EU-wide defense equipment market would bring. This includes an effort to identify research and technology development priorities for security objectives.

In this document, the Commission also called for a more coherent effort for European advanced research relating to global security. It proposed the launch of a preparatory project over a pilot phase of three years, implemented by the Commission with national research centers and industry, that would develop certain advanced technology aspects that would be particularly useful in carrying out Petersberg tasks. This approach would look at the best way for effective cooperation between national advanced research programs in the field of global security.

---

[5] Commission of the European Communities, *European Defence: Industrial and Market Issues. Towards an EU Defence Equipment Policy, Communication from the Commission*, COM(2003)113 final, Brussels, 11 March 2003.

After the European Council and the European Parliament express their support, it is expected that such a project will begin in 2004 with an initial annual budget of 15m Euros and a focus on a small number of selected areas of advanced technology, likely to be picked from issues such as container shipping security, advanced transport technologies, space, command and control, biological weapon detection, nanotechnology, biometrics, IT security and critical infrastructure protection. In order to succeed in pushing European advanced defense research to a new, practically adequate level, it would be of critical importance that this initiative is informed and driven by an awareness of the central technological and organizational challenges for European security and defense forces (i.e., transformation in line with the information-technology revolution, jointness and the shift to network-enabled capabilities).

New Approaches to Linking Strategy, Requirements, and Technology. Clearly, for picking the relevant projects and technologies in a forward-looking effort to prepare for requirements of the mid- and long-term future, there will be a need to draw on a new quality of advice that combines the strategic assessment of political, economic and military factors with a thorough awareness of technology; this effort should be similar in function to the Defense Science Board in the United States. Remarkably, the Commission's communication on defense industry explicitly mentions the desirability, in the longer term, of establishing a "European DARPA" that would provide impetus and a framework for advanced security and defense research with a view to leap-frog, follow-on generation technologies, as a complement to the more short-term research and development interest to pursued in national and European armaments agencies.

The "European Only" Dimension. This new European approach, as currently designed, would exclude non-EU partner countries, and only European industry would be allowed to participate. It is, however, pursued with the acute awareness that a successful European defense industry must form part of a larger market. With a strengthened defense-related R&D base that would support increased capabilities for future requirements, European countries and companies should become better able to gain access and respect in the United States, they would be better able to offer attractive skills and products.

The dominant motivation is to strengthen Europe vis-à-vis the United States, but this is not meant in the sense of rivalry and detachment. If properly accompanied by political leadership on both sides of the Atlantic, the development could, in the end, pave the way to the creation of a Transatlantic marketplace for defense goods and services that would provide EU firms with improved access to U.S. procurement, ease investment and permit the sharing of technology in this heavily protected sector. It is an open question, however, to which degree the desired strengthening of the European defense-research base and the European defense-industrial market will indeed be able to serve as leverage vis-à-vis the United States in this sense. One increasingly relevant argument for development in this direction, pointing to the U.S. interest in capable allies equipped to act responsibly and effectively on their own, results from the need for improved interoperability and doctrinal compatibility among allies at a time of rapid technological change and force transformation, and with major operations invariably conducted in multilateral coalitions. For the time being, the apparent strong political resistance against such an opening within the United States, notably in Congress, and the memory of shabby treatment of European allies by the United States in the past, greatly limits the degree of optimism on the European side about the realistic chances of a joint and networked transatlantic approach to defense technology in the foreseeable future.

Test Case: A Coherent European Space Policy

A number of factors, including the obvious "multiple use" nature of many space technologies and applications, the considerable funds earmarked for space-related research under the 6th Framework Programme, the cooperation agreement concluded by the EU with ESA, the White Paper on space with its broad agenda for action, and the prospect that the future EU constitution will include space among the shared competences of the EU, have all worked together to turn space into the dominant application area, for the time being, where the new approaches discussed in this paper are being implemented. The White Paper on space clarifies that while there will be an EU space policy, there will not be a single European space budget or only a virtual one. Space policy is not pursued as a goal in itself but because of the enabling role of space across the spectrum of EU policies. Space-related funds will be made available to those individual policy areas, while at the same time ensuring that their use makes the best possible contribution to a coherent space policy. With a view to the close connection between space assets and security, the White Paper states the decisive starting condition at the heart of any serious cooperative European effort in this field: "No single Member State will ever have the means to develop and support the full range of the necessary capabilities and better value for money could be achieved by various forms of cooperation at the EU level." The White Paper expresses a preference for multiple use approaches to future space assets in harmony with user requirements as determined on the European level. At the same time, it recognizes the need to better define and handle the interface between civil and military/security users and applications and to reconcile military and civil use of multiple-use assets, such as in the use of classified intelligence satellites for wider security purposes.

The two main projects under the common agenda pursued by the EU and ESA provide excellent examples of the multiple use approach: GALILEO, as a European constellation of satellites and ground stations for navigation, timing and positioning similar to the U.S.-owned GPS, planned to be operational in 2008; and Global Monitoring for the Environment and Security (GMES), as a framework for independent and permanent access to space-derived information and its broad use in support of EU policies on the environment, climate change, agriculture and fisheries, development, humanitarian aid, resource management, transport, regional development and "the quality of life and security of citizens", including foreign and security policy, early warning of crises, conflict prevention, crisis management and rapid emergency and disaster assessment. While GALILEO has been entirely defined as a civilian system, its public regulated service signals for exclusive government use are designed for security-sector users including armed forces. The "dual use" and hard security implications of GALILEO have only quite recently become an object of political discussion, triggered by the U.S. expression of security concerns over interference of GALILEO with next-generation GPS signals. National defense ministries have been reluctant to express interest in GALILEO, both for lack of a requirement and fear of draining scarce resources. So far only the French Ministry of Defense has tentatively offered to contribute a small amount of funding to GALILEO for inclusion of a more jam-resistant PRS signal.

GALILEO is the first major public/private partnership undertaken at EU level. It is expected to provide key lessons for the best approach to public/private finance in future projects. Some question marks remain behind the underlying business model that relies on large expected commercial revenues and major private investment. For the moment, however, GALILEO (with start-up costs of more than 3bn euro until 2008) serves above all as a necessary cash cow for a European space industry left without public contracts for too long and suffering from a steep downturn in commercial space markets. With the decision to open GALILEO to Chinese investment, formally agreed at the EU-China summit meeting in Beijing in October 2003, the project has both gained in credibility and assumed a new strategic dimension.

The current GMES Action Plan, unlike the initial approach, explicitly includes the objective of contributing to ESDP and the implementation of the Petersburg tasks, although the military dimension is currently not yet considered. The key notion is that Europe's ability to play a positive and effective role in conflict prevention and crisis management and thus contribute to the security of European citizens depends on its ability to draw on timely, accurate and reliable information. By establishing a focused, policy-driven framework for coordinating relevant R&D efforts, GMES is likely to prove helpful for strengthening Europe's capabilities in such areas as imagery data collection and exploitation, time-critical operational information support for decision-makers, and critical infrastructure protection. GMES is still in the early stages. The ambition of the approach lies, to a good part, in its ambiguity. Its security aspects may still meet political reservations and blockage from national governments. Some of the involved research ministries are adamant to keep their work entirely separate from anything related to defense or intelligence.

Perspectives: The EU Research Efforts as "Tender Flowers"

The EU's involvement in security-related research is a young phenomenon and a moving target. Dynamic developments are under way. Prospects for an intensified common EU effort to better exploit multiple use research and technology for European security and defense, driven to some degree by the Commission and to some degree by member states, look promising. However, little of what has been announced, agreed and initiated is yet fully supported throughout the EU on the political level, and budgets are uncertain. The political debate in member states, the European Council and the European Parliament still in its early stages. What one can see at the moment are simply some pretty, tender flowers growing in the shade. Most experts familiar with this development agree, however, that it is likely to be crucial for the future of European defense and the EU's role in the world, including the Transatlantic relationship.

Success or failure of the new multiple use research agenda is closely linked to the future success and ambition of the ESDP. Here, the recent endorsement of the first official EU security strategy at the European Council meeting in Brussels in December 2003 signals a milestone, as does the agreement reached at the same time, with NATO and U.S. blessing, on establishing a multilateral planning capacity for conducting possible future civil and military operations under the EU heading. To date, however, no long-term plan exists in Europe for the collective development of future military requirements and resulting research priorities on the European level. There is no institution where national projections are being actively harmonized. Available guidance on the substance of Europe's security-related research agenda is therefore very limited.

In a significant attempt to work towards overcoming this deficiency, the EU's white paper on space has launched a dedicated working group that is to present, by the end of 2004, its assessment of core questions faced in the development of a European multiple use space policy that contributes to security and defense:

- What are the space-related requirements of European policies in the field of security?
- What is the role of the future European armaments agency with respect to space? How do EU and ESA work together?
- What is the proper data access policy for multiple use imagery?
- What are the perspectives for the satellite center in Torrejón?

The group involved in this important effort includes the EU and its member states, ESA, and civil and military space users.

In sum, for sustained success, it will be crucial that a full, required understanding of the political, strategic, economic, and technological environment be inserted in the European research decision-making process. This can be helped considerably by active involvement from the private sector and third countries. The main test will be, however, whether the national defense and research establishments and defense ministries of major EU countries will realize in due time that they can indeed not deliver the required security and defense capabilities of the future without the kind of innovative European cooperation and integration approach that has now been put on track.

Chapter 6
European Defense Research and Technology (R&T) Cooperation: A Work In Progress

Andrew D. James

The character of European R&T cooperation is changing as European governments explore new means to address the well-known challenges that confront them. After languishing for more than a decade, the idea of a European armaments agency is back on the political agenda and research is included amongst its tasks. The European Commission is making its first formal foray into defense research with the launch of a pilot program of advanced research on security-related issues. There are moves to promote closer synergies between Europe's defense and civil research efforts. These are the latest in a series of developments since the late 1990s that have sought to address Europe's growing challenges in the field of defense research and technology. These challenges are easy to identify: finding the resources to allow more spending on defense R&T, developing means to improve the effectiveness of that spending, and promoting European R&T cooperation.

As discussed in detail below, a step-change in European defense R&T funding is unlikely in the current political and economic environment. However, European cooperative R&T is likely to broaden and deepen as those resource constraints force Europeans to work together more closely. Closer R&T cooperation alone may not lead to a step-change in European capabilities but it is a move in the right direction. Such developments will have important implications for future military capabilities and defense industrial competitiveness in Europe. Growing European aspirations towards a European Defense and Security Policy depend on the development of appropriate defense industrial capabilities. The transformation of European armed forces to support the Helsinki Headline Goal and their NATO commitments require ongoing investment in R&T. The widening R&T gap between the United States and Europe is making interoperability increasingly difficult. At the same time, R&T investment is a critical factor in the future competitiveness of the European defense industrial base.

Europe's R&T Challenges

There is little new in the defense R&T challenges that Europe faces today. Since the mid-1960s, European governments have sought to sustain some degree of defense industrial and technological autonomy against the massive R&T investments of the United States. The cost to NATO of defense R&T duplication was a central theme of the 1975 Callaghan Report.[1] Efforts to promote European defense R&T cooperation have been on the agenda for the last 30 years. Europe's R&T challenges are easy to identify but far more difficult to tackle.

Europe's first challenge is to spend more on defense R&T. In 2001, the U.S. government spent four and a half times as much on defense R&T as all the European Union countries combined.[2] The huge hike in US defense R&T spending to $58.6 billion in FY2003, combined with flat or declining European defense R&T spending, means that the United States is now spending almost six and a half times more than Europe. At the same time, R&D expenditure per head is considerably higher in the United States than it is in the European Union: $28,000 per head in the US compared with $19,000 in the UK, $12,000 in France and $5,000 in Germany.[3] Within Europe, spending on defense R&D is very unequal. In the WEAG, the six largest defense industrial countries account for 97 percent of R&D expenditure. The United Kingdom and France account for 77 percent of R&D expenditure.[4] At the same time, those governments depend upon company funded R&D to a greater extent than the United States with some European defense industrialists estimating that company funded R&D accounts for one-third of European defense R&D spending.

Europe's second challenge is to spend more efficiently and effectively. In many cases, individual national defense R&T systems in

---

[1] T. Callaghan, *US-European Economic Cooperation in Military and Civil Technology*, Center for Strategic & International Studies, Georgetown University, 1975.

[2] The International Institute for Strategic Studies, *The Military Balance 2001-2002* (London: Oxford University Press for the International Institute for Strategic Studies, 2001).

[3] G. Jordan, "Does there have to be a Trans-Atlantic Defence Technology Gap?" (presentation by the Science & Technology Director, Ministry of Defence, UK).

[4] The Western European Armaments Group (WEAG) has 19 full members: Austria, Belgium, the Czech Republic, Denmark, Finland, France, Germany, Greece, Hungary, Italy, Luxembourg, the Netherlands, Norway, Poland, Portugal, Spain, Sweden, Turkey and the United Kingdom.

Europe are very efficient and have become very effective at generating substantial scientific and technological breakthroughs from relatively limited funding (Sweden is a good example). However, when we view Europe as a whole national stove piping means that defense R&T spending is less efficiently used than that of the United States. Limited defense R&T resources are dispersed across national programs leading to considerable overlaps in research efforts and widespread duplication of laboratories and testing facilities.

This leads to Europe's third challenge—the need to cooperate more closely on defense R&T. The volume of cooperative R&T remains very limited with international funding accounting for no more than 5% of total funding allocated to defense R&T in Europe.[5] The UK has been one of the most vocal advocates of closer R&T cooperation but its expenditure on cooperative technology programs represents less than 8.5 per cent of its defense R&T spending.[6] This limited cooperative R&T activity is despite a widespread recognition of its benefits. A UK study found that international cooperation on R&T had a gearing effect that generated a rate of return to the UK government of 4.8 pounds (Sterling) for every one pound invested.[7] R&T collaboration in early phases of the acquisition cycle is seen as key to promoting equipment collaboration downstream. R&T cooperation can also have significant benefits in terms of promoting interoperability.

The History of European R&T Cooperation

Since the 1970s, growing concerns about the affordability of future development programs have led to attempts to share R&T costs through closer cooperation although, the volume of this activity has remained rather limited.

---

[5] Assembly of the Western European Union, *The Gap in Defence Research and Technology between Europe and the United States* (report submitted on behalf of the Technological and Aerospace Committee by Mr. Arnau Navarro (Rapporteur), Paris, France, 6 December 2000).

[6] National Audit Office, *Maximising the Benefits of Defence Equipment Cooperation* (report by the Comptroller and Auditor General, HC 300, Session 2000-2001, London, England, 16 March 2001).

[7] National Audit Office, *Maximising the Benefits of Defence Equipment Cooperation*.

IEPG

In 1975, the Callaghan Report argued that NATO was characterized by massive inefficiencies reflected in duplication of R&T, short production runs failing to exploit scale economies, and duplication of logistics support. Callaghan recommended standardization with defense industry rationalization and specialization throughout NATO and the extension of NATO cooperation to embrace civil technology.[8] In 1976, the Defense Ministers of the European NATO nations (except Iceland) established a forum for armaments cooperation, the Independent European Program Group (IEPG). The intention of IEPG was to strengthen the European pillar of NATO and improve armaments cooperation by consolidating the European defense industrial base and promoting economies of scale through cooperative armaments programs. In 1986, the Vredeling Report of the IEPG study group made a series of proposals aimed at improving the competitiveness of the European defense equipment industry that included the creation of a common European defense research program to strengthen the European technology base.[9] A loose set of Cooperative Technology Projects was established under the IEPG and in 1989 IEPG launched the European Cooperation for the Long term in Defense (EUCLID).[10] In practice, IEPG had a limited impact on European armaments and R&T cooperation not least because it lacked a political level to drive progress.[11]

WEAG/WEAO

In 1992, IEPG's functions were transferred to the Western European Union and WEAG was established. WEAG has traditionally been seen as the principal forum for armaments and R&T cooperation in Europe. WEAG's membership has grown from the original

---

[8] Callaghan, *US-European Economic Cooperation in Military and Civil Technology.*

[9] For a discussion of the Vredeling Report's recommendations, see T. Sandler and K. Hartley, *The Economics of Defense*, Cambridge Surveys of Economic Literature, Cambridge University Press, Cambridge, 1995.

[10] Assembly of the Western European Union, *Arms Cooperation in Europe: WEAG and EU Activities—Reply to the Annual Report of the Council* (report submitted on behalf of the Technological and Aerospace Committee by Mr. Piscitello (Rapporteur), Paris, France, 4 December 2002).

[11] F. Gevers, "Europe's Future Armaments Agency: is it Doomed to Repeat the Past?" *Jane's Defence Weekly*, 10 September 2003, p. 49.

thirteen European NATO-members who constituted the IEPG and today it has 19 full members. The objectives of the WEAG are the following: more efficient use of resources through, *inter alia*, increased harmonization of requirements; the opening up of national defense markets to cross-border competition; the strengthening of the European defense technological and industrial base; and, cooperation in research and development. These objectives have been pursued by three Panels: Panel I deals with cooperative equipment programs; Panel II is responsible for strengthening R&T cooperation; and, Panel III focuses on defense economic policy and armaments cooperation procedures.

The EUCLID program, involving industry and research institutes, is the main instrument for pursuing the WEAG's R&T activities. Within EUCLID, Common European Priority Areas (CEPAs) were set up in order to address relative shortfalls in technology compared to the United States. Further areas were added later and there are now 13 active CEPAs.[12] Some 82 specific Research and Technology Projects, which are part of the active CEPAs, have been initiated and 24 have already been completed. Current financing is around 100 million euros per year. Proposals from industrial consortia are taken into consideration through a mechanism called EUROFINDER established in 1996. The EUROFINDER process was established to allow multi-national consortia of companies to submit unsolicited proposals into the EUCLID program and it is run in parallel with the conventional, government-led approach to that program.[13] Cooperation between government research establishments has been facilitated through a government-to-government Memorandum of Understanding (MOU) called Technology Arrangements for Laboratories for Defence European Science (THALES) that was signed in 1996.

As a body to promote European armaments cooperation, the WEAG's results have been very modest not least because of its consensus-based decision making procedures and the conditions for col-

---

[12] E.A. van Hoek "European defense R&D and prospects for cooperation," (presentation by the Chairman of WEAG to a Symposium on "European Defense R&D: New Visions and Prospects for Cooperative Engagement," Center for Transatlantic Relations, Johns Hopkins University-SAIS, Washington DC, 6 June 2003).

[13] AECMA European Association of Aerospace Industries, *European Cooperation in Aerospace and Technology*, AECMA, Brussels, January 2001, pp. 24-26.

laboration whereby no member can be excluded from any collaboration.[14] Supporters of the WEAG argue that because states participate on an equal footing this avoids creating a "cartel" of powerful producers who can impose their aims and interests on other smaller countries and this ensures that the requirements of all states are taken into account and not merely the most powerful.[15] At the same time, there has been an absence of high-profile political support and an apparent lack of interest in WEAG's activities on the part of national authorities.[16] Indeed, in the absence of agreement at the highest political level, WEAG's work has inevitably been constrained by divergent national interests and procedures, and with technical and administrative matters that cause delay.

R&T has been the one area where WEAG has made progress not least because of the activities of the Western European Armaments Organization (WEAO). In 1996, the national governments agreed to establish the WEAO with a view to implementing, in particular, WEAG decisions in the research field. National governments gave the WEAO Research Cell the authority and necessary legal personality to place contracts and the Research Cell has become a useful and relatively effective tool for European research cooperation. Using the executive powers given it by national governments it has placed research and technology contracts worth 500 million euros involving 120 approved active research and technology projects.[17] The WEAO Research Cell has had some success but it has remained in practice a contracting agency with little power to coordinate research programs between national governments and the contracts that it has secured represent no more than 2.5 per cent of European military R&T spending.[18] Similarly, WEAG's Panel II has had limited powers to coordinate R&T programs. Under EUCLID rules, projects must be

---

[14] S. Törnqvist, "Demand Side Collaboration and Multi-National Procurement," *RUSI Journal*, Vol. 146, Issue 2, April 2001.

[15] Assembly of the Western European Union, *Arms Cooperation in Europe*.

[16] Assembly of the Western European Union, *Armaments Cooperation in the Future Construction of Defence in Europe* (report submitted on behalf of the Technological and Aerospace Committee by Mr. O'Hara (Rapporteur), Paris, France, 10 November 1999.

[17] Assembly of the Western European Union, *Arms Cooperation in Europe*.

[18] B. Schmitt, "The European Union and Armaments: Getting a Bigger Bang for the Euro," *Chaillot Paper No.63*, European Union Institute for Security Studies, Paris, France, August 2003.

notified to Panel II for formal approval. In practice, however, the Panel has almost no power to influence the choices made by its members on the establishment of projects.[19]

Cooperation between government research establishments

To address some of the problems of duplication of effort and facilities government research establishments have developed their own cooperative organizations and procedures, the Group for Aeronautical Research and Technology in Europe (GARTEUR), was established in 1971 based on a Memorandum of Understanding between the governments of France, Germany and the UK. Since that time, Italy, the Netherlands, Spain and Sweden have also become members. The objective of GARTEUR is to strengthen collaboration between European countries with major research capabilities and government funded programs through the exchange of scientific and technical information and efforts to avoid duplication of activities. GARTEUR also seeks to strengthen the competitiveness of the European aerospace industry by identifying technology gaps and performing joint research. The organization has sought to promote collaboration and harmonization of (European) government-funded research. GARTEUR has no central funding and no jointly staffed permanent secretariat or headquarters. The resources for joint research activities are made available by governments out of their national programs or by participating organizations on the basis of balanced contribution.[20]

European Research Establishments in Aeronautics (EREA) was established in 1994 as an association comprising the seven main European government aeronautics labs (CIRA of Italy; DERA of the UK; DLR of Germany; FFA of Sweden; INTA of Spain; NLR of the Netherlands; and, ONERA of France). The focus of EREA is civil and military aeronautics as well as space. There have been some efforts to enhance interdependence, integration, and pooling of resources as well as create centers of excellence. Some bi- and tri-lateral initiatives have been taken within EREA to pool resources within the existing boundaries of cooperation. ONERA (France) and DLR (Germany)

---

[19] Schmitt, *The European Union and Armaments*.
[20] AECMA European Association of Aerospace Industries, *European Cooperation in Aerospace and Technology*, AECMA, Brussels, January 2001.

merged their helicopter research work (and budgets) in December 1998; NLR (Netherlands), DLR and ONERA are working towards a joint organization (ATA: Aero Testing Alliance) to operate all their wind tunnels and over the years there have been discussions between NLR, DLR and ONERA on integrating major parts of their activities.[21] There is precedent for such a development. The Franco-German Saint Louis Research Institute has been in operation since the 1950s and for some years the UK, France, Germany and the Netherlands have been operating the European Transonic Windtunnel. However, there are considerable challenges: closer cooperation has to be accepted by the diverse national clients of each establishment and—although industry in Europe is increasingly transnational—other national clients still operate in a national environment. Equally, because the research establishments depend to a large extent on public funding, the pace of any developments is determined by national governments.[22]

The LOI Framework Agreement

The limitations of established R&T cooperation mechanisms (not least WEAG/WEAO) were the subject of attention during the Letter of Intent (LOI) process. On 6 July 1998, the Defense Ministers of France, Germany, Italy, Spain, Sweden, and the UK signed a Letter of Intent designed to facilitate defense industry restructuring in Europe. The LOI set up six specialist Working Groups to examine the main areas where the governments were committed to identifying concrete proposals to remove some of the barriers to restructuring. The Framework Agreement was signed in July 2000 and covers the following areas: security of supply; export procedures; security of classified information; treatment of technical information; research and technology; and harmonization of military requirements.[23] The Agreement directly addresses many of the principal impediments to more efficient and effective armaments cooperation and its sponsors hope that it will have significant benefits for the operation of both transnational defense companies (TDCs) and cooperative equipment

---

[21] AECMA, *European Cooperation in Aerospace and Technology*.
[22] AECMA, *European Cooperation in Aerospace and Technology*.
[23] The full title is the Framework Agreement concerning Measures to Facilitate the Restructuring and Operation of the European Defence Industry.

programs. The LOI Framework Agreement also includes provisions to foster coordination of joint research activities. The six nations have developed rules on the exchange of information on defense related R&T programs, they have begun the process of developing a common understanding of technology needs, procedures to allow for "closed projects," and competition as the preferred method for letting common R&T contracts and global return without requiring *juste retour* on an individual project basis. In addition, they have decided to develop a code of conduct to coordinate their relationships with TDCs and a Group of Research Directors has been set up to provide a focal point for the management of R&T cooperation.

EUROPA

The time and effort required to negotiate joint programs has often acted as a deterrent to cooperation in defense research. Traditionally, it could take up to two years to negotiate the necessary MOU and a year to negotiate the relevant technical annex.[24] Equally, in comparison with procurement programs, R&T projects tend to be more numerous, smaller (in financial terms) and very different in nature. R&T activities therefore need to be governed by specific, very flexible rules and procedures.[25]

The EUROPA MOU has sought to address some of these issues. EUROPA is an umbrella arrangement for R&T cooperation within the WEAO framework that provides that any two or more signatories can propose the creation of a European Research Grouping (ERG) to carry out bilateral and multilateral defense research and technology demonstration and testing of conventional defense related technologies. An ERG can be established to carry out a single large program or a number of individual projects. The first ERG was launched in late 2001 with 14 members and contains all the provisions necessary for the conduct of individual R&T projects. The first Technical Agreement (TA) setting up a cooperative project under ERG No.1 was signed by Italy and the UK in March of 2002 and offered considerable flexibility when compared with earlier R&T MOUs (e.g. THALES, EUCLID). Two or more ERG members can agree to take

---

[24] National Audit Office, *Maximising the Benefits of Defence Equipment Cooperation.*

[25] Schmitt, *The European Union and Armaments.*

part in projects without needing to seek permission from the whole group (closed projects); no automatic juste retour; work-share and/or cost-share will be decided on a case-by-case basis. The EUROPA MOU emerged in large part from the work on R&T issues undertaken as part of the LOI/Framework Agreement process and the EUROPA MoU was signed by WEAG ministers in May 2001. Since its provisions meet the expectations of the main arms producing countries the EUROPA MoU and ERG No.1 are likely to become one of the main instruments for the management of future European R&T projects. The nations have agreed that the EUROPA MoU and ERG No.1 will be used as the principle instruments for running specific projects.

The European Union

Significantly, European governments have chosen to pursue R&T cooperation outside the framework of the European Union. More generally, armaments questions have been left out of the European integration process not least because Article 296 of the Treaty on European Union (previously Article 223) excludes military goods from the common market and allows governments to exempt defense firms from European Union rules on mergers, monopolies and procurement. The role of the European Commission has been deliberately restricted by member states and has depended on the balance of the Commission's relationships with the Council and the Parliament.[26] However, the Member States have allowed the Commission to make rules that apply to certain activities related to armaments, particularly through competition regulations and in mergers and acquisitions involving defense-related companies and the control of exports of dual-use goods.[27] Moreover, the Commission is fully involved in the management of programs that may have consequences in the field of armaments, not least European space policy. These include the Galileo navigation satellite program, and the support for dual-use technologies under the Framework Programme for Research and Technological Development.

---

[26] C. Cornu, "Fortress Europe—Real or Virtual?," in Schmitt, B (ed.), "Between Cooperation and Competition: the Transatlantic Defence Market," *Chaillot Paper No. 44*, Western European Union Institute for Security Studies, Paris, France, 2001.

[27] Cornu, "Fortress Europe—Real or Virtual?".

The Sixth Framework for Research and Technological Development (FP6) is the European Union's main instrument for the funding of (civilian) research in Europe. Proposed by the Commission and adopted by the European Council and European Parliament in co-decision, it is open to all public and private entities, large or small. The overall budget for the four-year period of FP6 (2003-2006) is 17.5 billion euros, which represents 6 percent of the European Union public (civilian) research budget. There are no national quotas for FP6 funds that are allocated through competitive tendering. There are seven key thematic areas under FP6: genomics and biotechnology for health; information society technologies; nano-technologies and nano-sciences; aeronautics and space; food safety; sustainable development; and economic and social sciences. The main focus of FP6 is the creation of a European Research Area as a vision for the future of research in Europe. It aims at scientific excellence, improved competitiveness and innovation through the promotion of increased cooperation, greater complementarity and improved coordination between relevant actors at all levels.[28]

Aeronautics and space have been allocated EUR 1.075 billion under FP6 with the objectives of striving towards higher levels of technological excellence by consolidating and concentrating R&T development efforts. In the aeronautics field, the main thematic areas are: strengthening competitiveness by reducing development and direct aircraft operating costs as well as by improving passenger comfort; improving the environmental impact with regard to emissions and noise; improving aircraft safety and security; and, increasing operational capacity and safety of the air transport system. The main thematic areas for space are: Galileo—develop multi-sector systems, equipment and tools; GMES—stimulate evolution of satellite-based information services by developing new technologies (e.g. sensors, data and information models, services for global environment, land-use, desertification, and disaster management); and, satellite telecommunications—to be integrated with the wider area of telecommunications, notably terrestrial systems.[29]

---

[28] "New Framework Programme Launched: A Fact Sheet" (factsheet distributed by the Directorate-General for Research, European Commission, Brussels, Belgium).

[29] See http://www.cordis.lu/fp6/aerospace.htm.

Officially, the Framework Program funds only civilian projects but the Commission has long recognized (and accepted) that dual use technologies account for a substantial proportion of Framework Program funding and that the overlap is becoming greater. As one analyst observes:

> ... many aerospace companies have both civil and defence activities, and public R&T funding that helps them to remain competitive in civil markets is all the more important for them, as defence budgets continue to flat. Moreover, dual-use technologies have gradually been included in the Framework Programmes. GMES, secure telecommunication and safety of IT networks, for example, are all civil but security-related research programmes. They might well lead to applications that are of military interest, in particular in the IT and electronics sector, where military components and systems are increasingly developed on the basis of civil technologies.[30]

Lessons Learned

Three decades of European cooperation in the field of defense R&T have led to modest successes but they have also failed to resolve many key problems. The first critical lesson that we should have learned is that national governments continue to be reluctant to engage in cooperative R&T. Despite their statements of support for the concept, the reality is that cooperative R&T represents a mere 5 percent of all European defense R&T activity. There are several reasons why this is the case.[31]

Duplication and Its Consequences. First, the overlap between the various multilateral armaments cooperation organizations has led to competition for resources and the risk of duplicated research effort. Equally, European governments have been reluctant to share technical information where this is perceived to be of particular military or industrial advantage or because it might introduce vulnerability (such as through countermeasures). Thus, the more technically advanced countries have been reluctant to enter cooperative research in electronic warfare, sensor systems and signature control for fear of undermining their capabil-

---

[30] Schmitt, *The European Union and Armaments.*
[31] National Audit Office, *Maximising the Benefits of Defence Equipment Cooperation*, p. 30.

ity lead and industrial advantage in these areas. At the same time, the time that it has taken to negotiate joint programs has acted as a deterrent as has the fact that national governments have often found it difficult to integrate cooperative programs into their national R&T planning. All in all the consequence has been that international research cooperation—especially joint programs—has been seen as high risk.[32]

Difference Views on Technology Independence. The second lesson is that Europe's leading arms-producing countries have major differences in interests, placing limits on the progress possible even with these coalitions of the willing. Differences in industrial interests have always been important but equally critical have been differences in policy towards the objectives of European R&T. A key fault line between European governments has been the degree of technological independence or security of supply that is appropriate to Europe's circumstances. The UK has made its position clear in the MOD's science, technology and innovation policy. This emphasizes that UK defense technology policy is guided by Britain's desire to be able to make a distinctive, high quality contribution to multinational operations with equipment that is interoperable with the UK's most advanced allies and has a decisive technological edge over its opponents. This means that the UK accepts that some technology areas (not least those that underpin Network Centric Warfare) "will inevitably be led by the US."[33] In the UK's view, it is simply impossible to maintain a European capability in all areas. Europe should focus instead on developing Towers of Excellence in selected technology areas that, in turn, should be backed up by increased R&T spending and closer European cooperation. Equally, it will mean that Europe will have to accept dependency upon the United States in some technology areas. Other leading European defence industrial countries do not necessarily accept such views.

These policy issues may be a source of dispute between European governments but what has united them has been their belief in the primacy of national sovereignty in the armaments and R&T fields. The WEAG, the WEAO, OCCAR and the Framework Agreement have been the response of sovereign national governments to the

---

[32] National Audit Office, *Maximising the Benefits of Defence Equipment Cooperation*, p.30.
[33] Para.32, *Strategic Defence Review*, "Supporting Essay Three, The Impact of Technology", *op cit*, note 11 and *Defence Science and Innovation Strategy*, Ministry of Defence, 2001 available at http://www.mod.uk/issues/science_innovation/.

growing budgetary, industrial and technological pressures they have each faced. Armaments and R&T cooperation have been driven by national policies and interests rather than by any particular political will to build an "armaments Europe."[34] Time and again European governments have shown their preference for intergovernmental arrangements over European Union actions; OCCAR and the Framework Agreement represent forms of intergovernmental cooperation designed to protect national sovereignty.

To some extent, national governments have proven more willing to delegate powers to the bodies they have created. In a small way, the WEAO Research Cell is a good example of what can be achieved when executive powers are given to a body to implement the political will of national governments. Thus, the WEAO Research Cell has the power to conclude contracts on behalf of its member nations and, in doing so, administrative control has passed from national administrations to the international body. However, even in R&T, cooperation has suffered from a lack of support from member states and, more precisely, from the Research Cell's limited staff and restricted mandate (to administrative and contractual support).[35] Indeed, simply ceding administrative control is unlikely to help things if nations still keep the authority to release and direct funds. This has prompted some calls for the establishment of common budgets for research and technology in the context of a European armaments policy.[36] This exclusion of armaments questions from the European Union is having increasingly important consequences. The question of how Europe can best organize itself to exploit commercial technologies for defense applications has been the subject of growing debate and criticisms have mounted over what many see as the artificial divide between civil and military R&T in Europe. The exclusion of the European Commission from the defense field means that the Framework Program has been isolated from developments in European defense R&T through WEAG/WEAO, the EUCLID program and so forth. This divide makes little sense given that the Framework Program plays an important complementary role through its support for dual use technologies in aeronautics and space,

---

[34] Assembly of the Western European Union, *Arms Cooperation in Europe*.

[35] B. Schmitt, *The European Union and Armaments: Getting a Bigger Bang for the Euro*, Chaillot Paper No.63, August 2003, European Union Institute for Security Studies (Paris), p. 37.

[36] Assembly of the Western European Union, *Arms Cooperation in Europe*.

information and communication technologies, materials, and so forth. Indeed, it is a stark contrast to the organization of R&T in the United States where considerable efforts have been made to spin-in civil technologies for defense uses and promote civil-defense synergies.

The exclusion of armaments questions from the European Union has also had important implications for the development of the ESDP because it means that there has been no real linkage between developments at the European level and questions of requirements harmonization and equipment procurement. The need to develop the necessary military capabilities to support the Helsinki Headline Goals led to the launch of the European Capability Action Plan (ECAP) but there has been a growing recognition that this process lacks a mechanism to allow the conclusions of the ECAP process to be converted into coherent plans for capability development and, where necessary, new procurement programs.

A European Armaments, Research and Military Capabilities Agency

Addressing these challenges requires—in the words of one analyst—"a new institutional setting and a redefinition of competencies among the different actors in the field."[37] These realities have created a new political dynamic. The idea of a European armaments agency, having languished for more than a decade reemerged to take center stage, and European governments appear serious about developing closer cooperation in the field of armaments and R&T. The debates within the Convention on the Future of Europe led to the publication in May 2003 of a draft Constitutional Treaty that included a proposal to establish a European Armaments, Research and Military Capabilities Agency. The Agency would identify operational requirements, put forward measures to satisfy those requirements, contribute to identifying and implementing measures needed to strengthen the European defense industrial and technological base, participate in defining a European capabilities and armaments policy, and assist the Council in evaluating the improvement of military capabilities.[38]

---

[37] Schmitt, *The European Union and Armaments*.

[38] *Draft Constitution, Volume I*, CONV 724/03, Secretariat of the European Convention, Brussels, 26 May 2003.

The prospects for a European armaments agency had already been given a boost at the Le Touquet Summit between British Prime Minister Tony Blair and French President Jacques Chirac in February 2003. The Le Touquet Declaration on Strengthening European Cooperation in Security and Defense, stated that to support European capabilities goals and inter-governmental defense capabilities development, an acquisition agency could be established in the EU with the aim of promoting a comprehensive approach to defense capability development across all EU nations required for current and future ESDP missions. At the Thessaloniki European Council in June 2003, the European heads of state and government added a further sense of urgency to the discussions when they tasked:

> ... the appropriate bodies of the Council to undertake the necessary actions towards creating, in the course of 2004, an intergovernmental agency in the field of defence capabilities development, research, acquisition and armaments. This agency, which shall be subject to the Council's authority and open to participation by all member States, will aim at developing defence capabilities in the field of crisis of management, promoting and enhancing European armaments cooperation, strengthening the European defence industrial and technological base and creating a competitive European defence equipment market, as well as promoting, in liaison with the Community's research activities where appropriate, research aimed at leadership in strategic technologies for future defence and security capabilities, thereby strengthening Europe's industrial potential in this domain.[39]

The exact shape that the Agency will take was still the subject of debate in the Spring of 2004 and has already been the subject of considerable wrangling between national governments. However, the overall tasks of the Agency are clear. Article III-207 of the draft Constitutional Treaty states that the Agency:

> ... subject to the authority of Council of Ministers, shall have as its task to:

---

[39] Presidency Conclusions, Thessaloniki European Council (19 and 20 June 2003).

- contribute to identifying the Member States military capability objectives and evaluating observance of the capability commitments given by the Member States;

- promote harmonisation of operational needs and adoption of effective, compatible procurement methods;

- propose multilateral projects to fulfill the objectives in terms of military capabilities, ensure coordination of the programmes implemented by the Member States and management of specific cooperation programmes;

- support defence technology research, and coordinate and plan joint research activities and the study of technical solutions meeting future operational needs;

- contribute to identifying and, if necessary, implementing any useful measure for strengthening the industrial and technological base of the defence sector and for improving the effectiveness of military expenditure.

The history of European armaments cooperation suggests, above all, that grand political statements in support of a European armaments agency are one thing but turning them into reality may be quite another. Important issues will have to be addressed: the integration of existing armaments cooperation organizations; the membership of the Agency and the possibility of enhanced cooperation; the responsibilities of the Agency and the willingness of national governments to provide it with the necessary executive powers; and, concerns about European preference.[40] Of course, none of these issues is insurmountable given sufficient political will but they bring with them the prospect of potential delays in getting the Agency up and running.

What Role for the European Commission?

One issue that has emerged with surprising prominence is the role that the European Commission will play in these developments. The Commission has always been hamstrung in its efforts to broaden its

---

[40] For a detailed discussion of these issues, see A.D. James, "European Armaments Cooperation—Lessons for a Future Armaments Agency", forthcoming in The *International Spectator*.

competence into the armaments field by Article 296, the attitude of the European Council, and the hostility of some sections of the European Parliament. Thus, the publication in March 2003 of the Commission Communication *European Defence—Industrial & Market Issues* represents an important political step. The Commission has attempted this before only to be rebuffed by national governments but this time the political dynamic appears more favorable to the initiative.

The Commission Communication included a proposal for the Commission's first formal foray into defense-related research through an initiative to promote cooperation on advanced research in the field of global security and argued, "in a longer term the EU should consider the creation of a European DARPA". The Commission argued that this was a first step towards deriving greater benefits from national research programs through better coordination. The Commission announced its intention to launch a preparatory project that it would introduce with the Member States and industry to implement some specific aspects that would be particularly useful in carrying out Petersburg tasks. This preliminary operation would last no longer than three years and would cover a few carefully selected subjects of advanced technology together with specific accompanying measures.[41]

The budget for the preparatory action on advanced security research will be 65 million Euros over three years. The funding may be limited but this is an important political development as it marks the first time that the EU's research activities will focus specifically on defense-related issues. The Commission's aspirations in this area have been made plain by Research Commissioner Busquin who has called for closer synergy between research in the defense and civil areas and an end to the artificial division of European research activities. In a speech in September 2003, Commissioner Busquin made clear the Commission's intentions with regards the new Agency:

> [W]hen the Armaments Agency is born, we will have acquired an expertise in the field of security-related

---

[41] Commission of the European Communities, *European Defence—Industrial and Market Issues: Towards an EU Defence Equipment Policy*, Communication from the Commission to the Council, the European Parliament, the European Economic and Social Committee and the Committee of the Regions, COM(2003) 113 final, Brussels, 11 March 2003.

research which will be able to be integrated in this Agency, in coherence not only with the Community's research activities but also with national research policies which will coordinate more and more with each other within a Community framework.[42]

Commissioner Buquin also made clear that the preparatory action would make it possible to prepare the development of a security research program that the Commission could propose in 2005 or 2006.[43] The prospect of such a program being included in a Seventh Framework Program appears strong.

National governments have expressed their support for a new relationship between defense and civil research in the EU. The Brussels European Council in March 2003 invited an analysis of *"the role of defence R&T procurement in the context of the overall R&T activities in the Union*, including the possible creation by the Council of an inter-governmental defence capabilities development and acquisition agency."[44] However, turning the grand political statement of the European Council into practice is likely to generate tensions between Member States and the Commission. Equally, the effective "militarization" of even a small part of the Commission's research activities may provoke opposition from parts of the European Parliament. This is reflected in the fact that Commissioner Busquin has been at pains to point that "[w]e want to have a better synergy between civil and research and military research. But it is not envisaged to develop offensive weapons."[45]

Indeed, there is a danger that in walking this political tightrope, the original rationale for such a program may be lost. Much of the

---

[42] Author's translation of *"La Recherche dans la politique européenne de l'armement,"* Speech by Research Commissioner Philippe Busquin to the Kangaroo Group, European Parliament, Brussels, 9 September 2003 (http://europa.eu.int/comm/commissioners/busquin/research_talk/speeches/sp09092003fr.html).

[43] Author's translation of *"La Recherche dans la politique européenne de l'armement,"* Speech by Research Commissioner Philippe Busquin to the Kangaroo Group, European Parliament, Brussels, 9 September 2003 (http://europa.eu.int/comm/commissioners/busquin/research_talk/speeches/sp09092003fr.html).

[44] Council of the European Union, *Presidency Conclusions*, Brussels European Council 20 and 21 March 2003, para.35, p. 15.

[45] Author's translation of Research Commissioner Philippe Busquin, *"La Recherche dans la politique européenne de l'armement."*

rationale used by the Commission to date has talked about spillover effects from defense to civil technology development. Commissioner Busquin has said: "We pay a too high price in Europe for artificial separation between civil research and military research. . . . The example of the United States shows us that synergy between the two can be very profitable."[46] The Brussels European Council used a similar rationale stating, "The European Council recognises the role that defence and security related R&T could play in promoting leading-edge technologies and thereby stimulate innovation and competitiveness. . . ." In science policy terms, these assertions are questionable given that it is widely recognized that it is civil-origin technologies that have driven the innovation and growth of the last decade. From the point of view of Europe's defense R&T needs, there is a danger that the real requirement for increased European funding for defense and security related R&T may be muddied by talk of commercial spin-offs.

Given these political challenges the Commission has made remarkable progress. In October 2003, a "Group of Personalities" (GOP) was convened with the task of developing a vision for a future European Security Research Programme. The GOP comprised senior figures from the European aerospace, defense and scientific communities and was chaired by European Commissioners Philippe Busquin (Research) and Erkki Liikanen (Enterprise and Information Society). On 15 March 2004, the GOP presented its Report Research for a Secure Europe which argued the need for increased coordination in security research, outlining 12 recommendations for the future, including a minimum of €1 billion annually for security technology development. On 31 March 2004, a first call for proposals for projects and supporting activities under the new "Preparatory Action on the enhancement of the European industrial potential in the field of Security Research" was published. The Commission expects to fund six to eight projects under this first call, which has an indicative budget of million. Such initiatives have been given added political saliency by the Madrid train bombings.

---

[46] Author's translation of Research Commissioner Philippe Busquin, "La Recherche dans la politique européenne de l'armement".

Conclusions and Implications for Transatlantic Cooperation

European R&T cooperation is in the midst of significant change as European governments explore new means to address the well-known challenges that confront them. The idea of a European armaments agency, after languishing for more than a decade, is back on the political agenda and research is included among its tasks. The pilot program of advanced research on security-related issues marks the European Commission's first formal foray into defense research. There are moves to promote closer synergies between Europe's defense and civil research efforts. These are the latest in a series of developments since the late-1990s that have sought to address Europe's growing challenges in the field of defense research and technology. These challenges are easy to identify: finding the resources to allow more spending on defense R&T; developing means to improve the effectiveness of that spending; and, promoting European R&T cooperation. A step-change in European defense R&T funding is unlikely in the current political and economic environment but European cooperative R&T is likely to broaden and deepen as those resource constraints force Europeans to work more closely.

Closer R&T cooperation alone may not lead to a step-change in European capabilities but it is a move in the right direction and policy makers, and industrialists in the United States should unreservedly welcome it. Europe's failure to tackle its defense R&T challenges has had major implications for transatlantic relations. On the one hand, there is the problem of the growing technology gap and the challenge of interoperability. On the other, there is the question of economic competition and the perceived threat of U.S. dominance. By improving the effectiveness of European R&T efforts these initiatives ought to support the general thrust of the NATO transformation agenda and enhance Alliance interoperability. Stronger European R&T activity may also enhance transatlantic partnership opportunities with more technologically capable European companies opening up new mutually beneficial alliance opportunities. However, this very much depends on the U.S. industrial and political response. There are those in the United States who worry that a European armaments agency represents a step towards a policy of European preference and the creation of a "Fortress Europe". There are some in Europe who would like it to become so but governments outnumber them and companies who wish to create the conditions for a healthy and balanced transatlantic relationship. More importantly, access to U.S. technology will continue to be a major concern for Europe and U.S. reforms of export and technology controls remain a vital prerequisite, if we are to strengthen transatlantic technology exchanges

Chapter 7
International Defense R&D Cooperation: From Competition to True Cooperation—The Case of U.S.-Japan Defense R&D Cooperation in Transition

Dr. Masako Ikegami[1]

International Defense R&D Cooperation After the Cold War

With the end of the Cold War, the geopolitical and strategic conditions have evolved from a clear threat definition and assessment to an uncertain and more diffuse security posture. Clearly, this new situation is not favorable to the continued mobilization resources on a large scale for the development and production of advanced weapon systems against a single strategic threat. Currently, post-Cold War armed forces are required to deal with a range of missions other than armed combat; e.g., humanitarian and peace-keeping operations, rescue services, and anti-terrorism operations. Consequently, relatively limited defense resources now must be spread over a variety of weapon systems. In order to deal with the post-Cold War circumstances, international armaments cooperation became imperative. Key "drivers" that also foster the ongoing globalization of the defense industry include the 'internationalization' and 'commercialization' of military technology and weapon systems as well as the changing framework of post-Cold War security policy.

---

[1] Author's Note: The author would like to thank Richard Bitzinger, Gegg Rubinstein, Wayne Fujito, James Armington, Dr. Björn Hagelin, Dr. Manahiko Ueda for providing the author with their rich insights on the issue. Also gratefully acknowledged is useful materials and data provided by the Stockholm International Peace Research Institute (SIPRI), the Japanese Ministry of Economy and Industry (Divisions of Aerospace and Arms, and of Security Export Control), Defence Research Center (DRC), and Defence Production Committee of the Japan Federation of Economic Associations (*Keidanren*). The author is grateful to Jeffrey Bialos and Stuart Koehl, Johns Hopkins School of Advanced International Studies (SAIS) for the exclusive opportunity for the author to present a paper at the 6 June 2003 symposium.

International collaboration on developing new weapon systems is not a new phenomenon; there were many examples even in the early period of the Cold War, such as inter-European and U.S.-European collaboration. For European countries in particular, multinational collaboration was a matter of course given the limited scale of the resources and market of each country. In Europe, active intra-European restructuring of the defense industry began in the 1980s. However, in the post-Cold War era, the internationalisation of defense R&D is even becoming important for the United States, the world's outstanding investor in defense R&D. The more sophisticated weapon systems become, the higher the technical risks and development costs. In contrast to the Cold War period, the lack of clear threats[2] and political legitimacy has increased the difficulty of making excessively large-scale resource commitments exclusively for military purposes. (see **Table 1**).

Table 1. Government Defense R&D Expenditure in select countries, 1986-97, in million U.S.D at 1995 constant prices and exchange rates.

Country	1986	1989	1992	1993	1994	1995	1996	1997	1992-96
U.S.A	51,000	51,000	44,000	43,000	39,000	37,000	37,000	38,000	200,000
France	6,200	7,100	6,800	6,200	6,000	5,200	5,000	4,600	29,200
UK	5,400	4,100	5500	800	300	300	400	300	17,300
Germany	2,300	3,100	2,400	1,900	1,900	2,000	2,200	2,100	10,400
Japan	[820]	1,100	1,400	1,500	1,500	1,600	1,800	1,800	7,800
Italy	540	750	600	620	590	[560]	[680]		3,500
Sweden	660	680	690	650	500	570	570		2,980

Source: Eric Arnett, 'Military Research & Development,' SIPRI Yearbook 1999, p. 353.

Although nearly all of the industrial states prefer 'self-sufficiency' in armaments, defense industries have formed global networks to access foreign markets, technologies, and capital. The result is enhanced international collaboration for weapons development and production. International armament collaboration can take the form of co-production or co-development. Acquiring 'high-tech' weapon systems depends on the availability of two key resources—technology and money. According to Ethan Kapstein's categorization, there are four major types of acquisitions (see **Figure 1**).

---

[2] International terrorism is recognized as the new threat, but much in a diffused form. Thus, it does not necessarily lead to procurement of a large-scale weapon systems.

Figure 1. State Strategies for Acquiring Weapons
State Acquisition Preferences:
Autonomy > Co-develop > Co-produce > Import

		Financial Yes	Assets No
Technology	Y	Autonomy (Rafael)	Co-develop (Nunn Amendment)
Assets	N	Co-produce (F-16)	Import

Source: E. Kapstein 1992, p. 4.

According to Kapstein, "co-development and co-production are 'second-best' solutions that enable the state to maintain a domestic defense-industrial base and to share the financial and technical risks of weapons development with alliance partners." Due to restrained military expenditures after the Cold War, many countries take the "second-best" solutions. In his view, the 'internationalization' of defense R&D not only implies an increase in international defense R&D collaboration but also that international collaboration is becoming an imperative, due to the changing post-Cold War of economic, political and strategic conditions.[3]

Another trend, 'commercialization' of defense R&D implies that cutting edge civilian technology is increasingly utilized in weapons development and production as an important technological base and for cost savings in the post-Cold War era. Development and production of advanced weapon systems rely not only on government-sponsored defense R&D; *per se*, but also on the nation's industrial and technological base as a whole. In this context, dual use technology has become a key element of contemporary defense R&D; 'spin-on' has now supplemented 'spin-off'.[4] Further, dual use technology has become increasingly important in terms of cost saving for weapons development and production in the post-Cold War era. The pressure to reduce military expenditures military expenditures has led to an increasing reliance on commercially developed and produced components and sub-systems

---

[3] In this study, 'internationalization' denotes increasing cross-border transactions among national actors, while 'globalization' denotes more qualitative changes.

[4] In the 1980s, the Department of Defense (DoD) conducted major studies of 'critical technology,' that was important from the perspective of national security. The studies compared competitiveness and advantages between American and Japanese critical technologies, and identified some Japanese dual use technologies that would be of significance for improving U.S. weapon systems.

being incorporated into weapon systems rather than very expensive, custom designed and manufactured" "military standard" components.

The Case of U.S.-Japan Defense R&D Cooperation

The History of U.S.-Japan Cooperation in Defense R&D

In the case of Japan, the internationalization of the defense industry means almost exclusively armaments cooperation with the United States due to both the U.S.-Japan security treaty as well as Japan's strict restrictions on the export of arms and military-related technology. Although the Japanese defense industry had produced many major weapons systems under U.S. licenses for decades, substantial collaboration really began when the two countries initiated defense R&D cooperation in the 1980s. Given the 'internationalization' and 'commercialization' of defense R&D, the United States took the initiative to strengthen defense R&D collaboration with Japan from the early 1980s. In 1980, pursuant to a U.S. initiative, the United States and Japan established the System and Technology Forum (S&TF), for facilitating "open and substantive dialogue" and cooperation between the Japanese Defense Agency (JDA) and DoD in the R&D, production and procurement of military equipment. In 1983, Japan decided to "open the way for the transfer of its military technology to the United States as an exception to the Three Principles on Arms Exports and other regulations."[5] This momentum was spurred by a number of factors, including growing U.S. interest in Japanese dual use technology and sharing R&D costs, the Japanese government's political desire to strengthen U.S.-Japan security cooperation, and the Japanese defense industry's expectation of more major programs and defense business.[6]

The first U.S.-Japan cooperation in defense R&D involved joint research as part of the Strategic Defense Initiative (SDI) program. Responding to then U.S. Defense Secretary Weinberger's invitation addressed to U.S. allies in 1985 to engage in SDI research coopera-

---

[5] (*Defence of Japan* 2001: 182). Accordingly the MoU, agreements established the 'Exchange of Notes on the Transfer of Military Technologies to the United States' in 1983, and the 'Detailed Arrangements for Transfer of Military Technologies' in 1985. Since then Japan has provided the United States with 12 items of military technology including portable SAM (surface-to-air missile) technology and construction technology for U.S. naval vessels, etc. (*Defence of Japan* 2001: 182).

[6] For details about collaboration on SDI, FS-X/F-2 and TMD, see Ikegami-Andersson (1998) *Military Technology and U.S.-Japan Security Relations*, Uppsala & Stockholm.

tion, the Japanese defense industry joined an architecture study for SDI based on the U.S.-Japan government-to-government Memorandum of Understanding (MOU) signed in 1987.[7] The second major cooperative project was the FS-X/F-2 fighter support aircraft which turned out to be highly controversial because of technology transfer issues between Japan and the United States.

The third major project is the ongoing Ballistic Missile Defense (BMD) cooperation in the Navy Theater Wide Defense (NTW) program, through which JDA is conducting requirement analysis and design (RA&D) of four components: a nosecone to protect the infrared seeker from heat, a kinetic warhead, an infrared seeker to detect and follow target, and a second-stage rocket monitor of a three-stage missile.[8] These activities comprise a marginal part of the overall NTWD system. However in the context of two-decade history of U.S.-Japan defense R&D collaboration, they nevertheless are important. Japan joined a program early in the definition phase.[9] The JDA budgeted 3.708 billion yen in FY2001 and 6.937 billion yen in FY2002 as expenses for the design and trial manufacture of the four main components of the NTWD.[10] In addition to these major programs (SDI, FS-X/F-2, and the ongoing NTWD) the United States and Japan also have been conducting nine joint research programs for various components and sub-systems since 1992 (see **Table 2**).

---

[7] The cooperative project was a feasibility study called the Western Pacific Missile Defence Architecture (WESTPAC) with relatively small-scale funding from the DoD. Two groups of industrial consortia were designated for the study: One is the Mitsubishi Heavy Industry-led consortium which includes Mitsubishi Electronics, NEC, Hitachi, Fujitsu, Japan Radio Corporation, Mitsubishi Space Software, Mitsubishi Corporation, as well as SAIC of Alabama, Boeing Aerospace, Lockheed Missile System and Raytheon; the other consortium is led by LTV Aerospace & Defence and includes Kawasaki Heavy Industry and Analysis Pacific (*Armed Forces Journal International*, February 1989, and GAO, *SDI Program: Extent of Foreign Participation*, February 1990). Research funds for WESTPAC totaled $8.5 million over the four years (1988-1992) (Yamashita et .al., 1994, *TMD: Seniki Dando Misairu Boei*, Tokyo: TBS Britanica).

[8] *Defence of Japan* 1999: 84.

[9] "The NTW Program is currently in the Program Definition and Risk Reduction Phase of development. The navy intends to propose the two-phase approach...The first phase, Block I will address the current preponderant TBMD threat. NTW Block II will be treated as a major acquisition upgrade to the Block I Program. For the purpose of this study, the NTW Standard Missile-3 (SM-3) (Block II) was used as the baseline for analysis ... The Block II NTW system is not completely defined or fully funded" ('Summary of Report to Congress on Utility of Sea-based Assets to national Missile defence', BMDO, 1 June 1999, p. 9).

[10] (*Defence of Japan* 2002: 199)

Table 2. Items of Japan-U.S. Cooperative Research and Modification Projects

Project	Summary	Agreed Year
Ducted rocket engine	For the secondary combustion of solid rocket fuel	September 1992 concluded 1999
Advanced steel technology	Extra-high-strength steel for submarines	October 1995 concluded 2001
Fighting vehicle propulsion technology using ceramic materials	Diesel engines using ceramic materials	October 1995
Eye-safe laser radar	LIDAR systems using eye-safe frequencies	September 1996 concluded 2001
Ejection seat	Modification work to supplement combat aircraft	March 1998 concluded 2001
Advanced hybrid propulsion technology	Thrust-controllable propulsion devices made up of solid fuel and liquid oxidizers	May 1998
Shallow-sea water acoustic technology	Transmittance of sound waves in shallow sea regions	June 1999
Ballistic Missile Defence technology	Four components of Navy Theater Wide Defence (NTWD)	August 1999 March 2000
Low-vulnerability gunpowder for field artillery	Gunpowder that avoids unintentional secondary explosions	
Avionics aboard the follow-on aircraft to the P-3C	Onboard avionics of the MSDF's next fixed-wing maritime patrol aircraft (P-X) and the U.S. Navy's future Multi-purpose Maritime Aircraft (MMA) for better inter-operability	March 2002
Software radio	Basic technologies for software radio which enables primary radio functions through software	March 2002

Source: Defence of Japan 2002, The Japanese Defence Agency, Urban Connections Publisher, p. 197.

Financial Constraints on R&D Cooperation

The key to Japanese defense industry's internationalization continues to be advanced dual-use technology. Japan has a number of attractive dual-use technologies that are useful both in terms of improving precision and quality as well as achieving cost-savings through COTS products. Since the 1990s, Japanese defense expenditures, particularly that of equipment procurement, have largely remained flat. In the previous Mid-Term Defense Program (MTDP) (FY1996-2000), the expenditure for major weapons procurement declined by 1.8% (at 1995 constant price); for MTDP of FY 1991-1995, the net reduction was 6.2% (at 1990 constant price).[11] In the new MTDP (FY 2001-2005), the total budget for front-line equipment is 4.03 trillion Yen (at FY2000 price), as compared to the corresponding figure of the previous MTDP (1995-2000) budget of 4.07 trillion Yen (in 1995 prices) (*Defence of Japan* 2001: 104).

Figure 2. Declining Defense R&D Budget, 1996-2000
(in hundred million Yen)*[12]

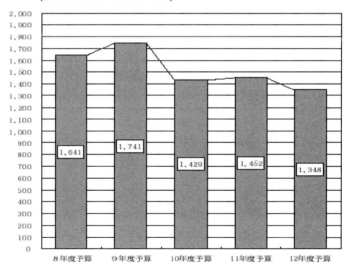

---

[11] (*Defence of Japan* 2000: 86)

[12] *Source*: JDA Report 'Basic Guidance for maintaining and developing technological infrastructure of the Japanese defence industry' (in Japanese), November 2000, (http://www.jda.go.jp)

Figure 3. Declining Budget for Frontline Equipment Procurement Contract base (1989-2000)

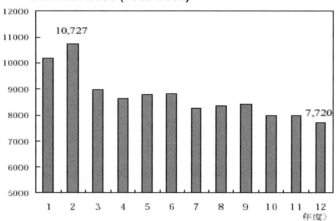

Figures in hundred million Yen.[13]

Figure 4. Declining Budget for Frontline Equipment Procurement as ratio of FY2000 Peak-time value (tanks/ammunitions, vessels, aircraft, guided missiles)

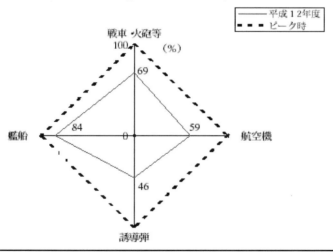

Red line for FY 2000; Black dots for the peak-time*[14]

---

[13] *Source*: JDA Report 'Basic Guidance for maintaining and developing technological infrastructure of the Japanese defence industry' (in Japanese), November 2000, (http://www.jda.go.jp)

[14] *Source*: JDA Report 'Basic Guidance for maintaining and developing technological infrastructure of the Japanese defence industry' (op.cit.)

Political and Legal Constraints on U.S.-Japan Defense R&D Cooperation

The scope of Japanese defense industry's international cooperation is significantly constrained by the Three Principles of Arms Exports. To date, except for the United States (based on the 1983 and 1985 MOUs), Japan has not transferred its military technology to foreign countries. Although there is a growing interest in Japan having more technological exchanges with other foreign countries, there has so far been very limited substantive engagements. The lack of Japanese government cooperation on defense with third countries in turn limits the incentives for the Japanense industry to engage in international cooperation. In the absence of such international demand, the Japanese industry is not in a position to profit from international defense cooperation or defense technology exchange. In an effort to encourage the Japanese industry's participation in U.S.-Japan defense R&D cooperation, the U.S. and Japanese governments are considering the partial relaxation of the Three Principles.[15]

Further the Japanese Constitutional prohibition on "collective defense" is one of the major barriers to Japanese defense industrial cooperation with the United States on the NTWD project. If or when NTWD becomes operational, much closer interoperability between the U.S. and Japanese C3I is needed in order to make the system work most effectively; this will touch upon 'collective security,' which is currently prohibited under the Japanese Constitution.[16] The limited Japanese budget for defense R&D also fuels the JDA's reluctance to contribute more funding to proof-of-principle tests for the NTWD project.[17] Thus, due to the political, legal, and financial constraints, Japan's defense industry remains relatively unengagedfrom the internationalization of defense industry.

---

[15] Interview with Mr. Gregg Rubinstein, GAR Associates, Washington DC, May 2000.

[16] However, the 'collective defence' barrier is being almost virtually lifted; Japan decided in October 2001 to dispatch its naval escort vessels to the Indian Ocean to provide logistic support in the rear for the U.S. anti-terrorism operations in Afghanistan (Special Act on anti-terrorism measures, approved by the Diet on 29 October 2001). This is the first time that the SDF will provide support logistics for U.S. forces in combat.

[17] The Japanese side's reluctance to contribute more for testing has been rather disappointing for the U.S. (interview with a former Pentagon official, currently defence-related manager, DC, May 2001).

Impediments to Further Cooperation: "Kokusanka" and Protectionism

Due to the Three Arms Export Principles, Japan's defense industry is relatively closed, with the JDA as the only buyer. Defense procurement still is based on 'cost plus'; this means the defense industry fares better when its costs are higher. Thus, the defense industry has few incentives to work seriously on cost reduction.[18] Not surprisingly, in light of defense budget constraints, the Japanese defense industry has undergone a similar consolidation to that experienced in Europe and the United States. In fact, most Japanese companies have very little share in defense production,; which is very small in Japan (0.62% in 1999 and 0.58% in 2000) relative to total industrial production.[19] Given the financial constraints, many defense-related companies are shifting to civilian production. Recent major (and rather exceptional) mergers, acquisitions and resource-sharing agreements included the following: Ishikawajima-Harima Heavy Industries (IHI) and Sumitomo Heavy Industries established an equal partnership in October 1995 in an engineering company for R&D, design and life-cycle support of naval vessels, the Marine United;[20] Mitsui Engineering & Shipbuilding Co. and Hitachi Zosen Corp. signed a resource-sharing agreement;[21] IHI acquired the aerospace division of Nissan Motor Co. in 2000, which is the "first ever take-over of a weapon producer in Japan";[22] and NEC and Toshiba Corp. merged their satellite manufacturing operations into a 50/50 partnership to make common use of their respective facilities.[23]

---

[18] Interview with an anonymous official of a Japanese defence-related trading company, November 2001.

[19] *Defence of Japan* 2002, p. 416.

[20] (*Defence News*, 8-14 May 1995)

[21] (*Defence News*, 26 Jun-2 July 1995)

[22] While HIH tries to streamline its aerospace and military operations as one of its future core business, Nissan has faced a serious re-engineering and re-focus on its automobile business, since the M-5 rocket which the company manufactured failed in February resulting in the loss of a $105 million Astro-E X-ray astronomy rocket; HIH was a contractor for the now-cancelled J-1 and H-2 rocket programmes. However, the HIH-Nissan deal is likely to remain an isolated case, and will not trigger a rush of mergers of defence suppliers, largely because Japanese weapon builders are far less dependent on that business than their American counterparts" (ibid. and *Industry*, no. 14,430, 16 Feb. 2000).

[23] (*Aviation Week & Space Technology*, 13 March 2000)

While there is a growing trend in international defense cooperation, there still is traditional *Kokusanka* (indigenous R&D and production) orientation among Japanese defense companies. For instance, the new Mid-Term Defense Program (MTDP) for FY2001-2005 includes two major *Kokusanka* projects, including development of "the follow-on aircraft to the P-3C fixed-wing maritime patrol aircraft and the C-1 transport aircraft."[24] These projects are primarily for maintaining the Japanese aerospace industry's infrastructure; as for the P-3C follow-on (P-X), Japan's initial focus was on indigenously developing the aircraft itself rather than the avionics sensors and electronics—which are to be the core function of the system.[25] In this sense, these projects may derive from industrial considerations rather than military requirements.

Another major defense-related project is the eventual launch four reconnaissance satellites by the indigenously developed H-IIA rocket. Although the Japanese industrial association is aware of the limited capability of the indigenous version of the satellites, these efforts will likely continue, since technological competitiveness is regarded as vital bargaining power. The civilian aerospace industry (e.g. satellite, rockets, ground facilities) continues to be strategically important due to the great spin-off effects expected. While Japanese aerospace industry was heavily subsidized by the government, it is currently transitioning toward operating on a more commercial basis.

In sum, due to political and legal contraints, the strategically important Japanese defense industry still remains national, indigenization-oriented, and 'protectionist' in orientation.. Nonetheless, globalization of the defense industry is irresistible, and U.S.-Japan armaments cooperation is gradually evolving.

U.S.-Japan Defense R&D Cooperation: A Work in Progress

U.S.-Japan defense industrial cooperation still remains limited in nature, shaped more by industrial considerations than considerations of common security or doctrine. Unlike the case of

---

[24] (*Defence of Japan* 2001: 100-1)

[25] (*Defence News*, 24 April 2000). However, there is possibility for the U.S. and Japan to co-develop avionics and electronics for future Maritime Patrol Aircraft; if so, it would be "the first real example of a requirements dialogue between the U.S. and Japan" (CPAS-UI Seminar by Gregg Rubinstein, Gar Associates, at the Swedish Institute of International Affairs, Stockholm, 27 September 2001).

NATO, there is no institutional mechanism to support armaments cooperation by way of coordinating requirements or common doctrines.[26] Compared to Transatlantic defense cooperation, U.S.-Japan defense industrial cooperation lags behind in various respects. Although Japan is included in the Defense Trade Security Initiatives (DTSI) initiated by the Clinton Administration, Japan seems to be more like an 'honorary' rather than an active member.[27] Its political and legal constraints and limited institutions for the international cooperation (e.g. the protection of military information and intellectual property) may be the cause of the lag. However, since Japan is important for the United States in terms of buying U.S. weapon systems and for U.S. strategy, Japan will continue to be engaged, albeit in a passive way, in defense industry globalization through the United States.

In spite of various constraints and impediments, U.S.-Japan armaments cooperation is gradually evolving after all. Given the financial constraints and increasing international competition, JDA is pursuing major procurement and acquisition reforms for "strengthening market principles and reducing life-cycle costs to ensure transparency and fairness in procurement of equipment and services and to establish efficient systems of procurement and supply."[28] Accordingly, JDA completed a major re-organization of the Central Procurement Office in January 2001.[29] For reducing equipment life-cycle costs, JDA also has introduced a new concept, the continuous acquisition and life-cycle support system (CALS), which focuses on cost-effectiveness and internationalized-level price setting for domestic defense procurement. JDA also introduced a new "Guideline for the Implementation

---

[26] "The dialogue on common interests in defence requirements and cooperation on future acquisitions evident in NATO has been almost absent between the U.S. and Japan" (CPAS-UI Seminar by Gregg Rubinstein, op.cit.)

[27] To see DTSI enacted toward Japan, it may require concrete programs such as Japan's licensed production of the Aegis system and possible co-development of maritime patrol aircraft. DTSI would then promote industry-to-industry cooperation.

[28] (*Defence of Japan 2001*: 159)

[29] Functions of the JDA's Central Procurement Office are now separated to Bureau of Administration of the Internal Bureau (for cost accounting), Central Contract Office (for contract), and Defence Procurement Council. Also JDA's procurement is now shifting to more open bids both in terms of the number of contracts and contracts in price (*Defence of Japan* 2001, p. 160).

of Defence R&D" in 2002 in order to broaden its options for international defense technological cooperation.[30]

Further, in order to deepen U.S.-Japan armaments cooperation, the U.S.-Japan Industry Forum on Security Cooperation ("IFSEC") was set up in 1996 as a defense industry initiative. IFSEC aims to broaden and promote industry-to-industry dialogue and cooperation between the two countries. IFSEC submitted a "U.S.-Japan joint statement on mutual interest" to their respective government with five recommendations: 1) expansion of dialogue on equipment cooperation, including the use of commercial items; 2) a positive attitude toward technology transfer; 3) more flexible application of Japanese export control policies concerning U.S.-Japan cooperation in defense equipment programs; 4) intellectual property protection; and 5) the need to address the impact of 'Buy American' provisions.[31] IFSEC also considers the DTSI as instrumental for facilitating industry discussion in specified mission or product areas and encouraging the use of an 'integrated product teams' (government, industry, customer) in international programs, DTSI should facilitate information disclosure and the processing of export licenses.[32]

As the two countries deal with these agenda, U.S.-Japan armaments cooperation will inevitably evolve toward more substantial collaboration such as requirements dialogues, cooperative development, and acquisition.[33]

---

[30] "So far, Japan's technological exchange with foreign countries has been constrained except for the United States, due to the Three Arms Export Control and secret-keeping, etc. However, given changes caused by globalization in and outside of Japan, Japan may well promote technological exchange with other countries even within the existing framework. Particularly it is useful for Japan to exchange technology with the countries that belong to U.S. alliances, those who have similar technological level and R&D milieu and those who have highly advanced military technology. To be concrete, Japan will consider possible measure for technological exchange with, for instance, France with whom Japan has an exchange since 2000, Republic of Korea, and the UK who has proposed technological exchange" [author's translation], 'Kenkyu-Kaihatsu Jisshi ni kakawaru Gaidorain [Guidelines for R&D Implementation]', Tokyo, JDA, 27 June 2001.

[31] 'IFSEC Joint Report 2003 –Revised U.S.-Japan Statement of Mutual Interest–', U.S.-Japan System & Technology Forum, February 2003, Keidanren/NDIA.

[32] (ibid., p. 3-4)

[33] (Gregg Rubinstein, 'Evolution of U.S.-Japan Armaments Cooperation' September 2001)

Figure 5. 'Evolution of U.S.-Japan Armaments Cooperation'

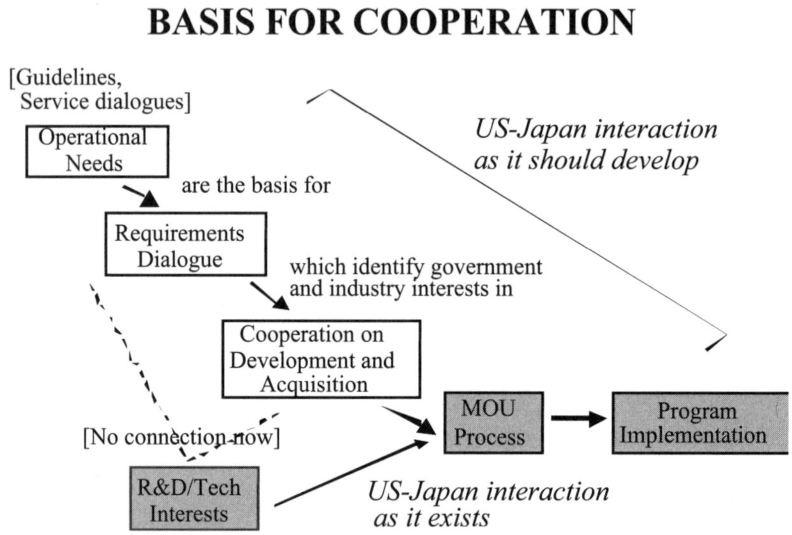

BASIS FOR COOPERATION

[Guidelines, Service dialogues]

Operational Needs are the basis for

Requirements Dialogue which identify government and industry interests in

Cooperation on Development and Acquisition

[No connection now]

R&D/Tech Interests

MOU Process → Program Implementation

US-Japan interaction as it should develop

US-Japan interaction as it exists

Source: Gregg Rubinstein, GAR Associates, September 2001.

New Trends in International Defense R&D Cooperation

Consolidation & Privfatization

Combined military equipment expenditures of all NATO countries dropped by 40% in real terms from the peak level in 1987 to 2001—by 43% in the United States and by 35% in NATO Europe, although with great variation between countries.[34] In 2001, the United States accounted for 28% of global arms transfers, but its delivery fell by 65% from 1998 to 2001, caused mainly by a drop in deliveries of combat aircraft to major recipients.[35] Estimates of national arms sales—used as an approximation of arms production—for the seven largest arms-producing countries in Western Europe also showed significant declines between 1990 and 2000 (see **Table 3**).

---

[34] (*SIPRI Yearbook* 2002: 324)

[35] (*SIPRI Yearbook* 2002: 374)

Table 3. Changes in National Arms Sales and Arms Exports, Western Europe, 1990-2000 (%)

	UK	France	Germany	Italy	Netherlands	Sweden	Spain
Arms sales	−21	−47	—	—	—	+19	−6
Arms exports	−27	−61	−29	−63	—	+4	—

*Changes are based on figures in million U.S.D at 2000 prices and exchange rates.
Source: SIPRI Yearbook 2002, 'Military Spending and Armaments, 2001,' p. 324.

The decline in demand for military equipment and soaring R&D costs reinforced the concentration and internationalization of defense production. According to a SIPRI study, the rate of concentration has increased significantly among the 100 largest arms-producing companies on the SIPRI Top 100 list (see **Table 4**). The SIPRI study points out that the concentration ratio in defense industry has resulted in a rate almost similar to that of non-military markets, which will possibly raise difficult political issues; in their procurement processes countries will confront large arms-producing companies with strong market power.[36] According to the study, the increased concentration has also resulted in an increase in the size of the largest arms-producing companies, both in relation to other companies and in relation to the total procurement budgets of domestic governments.[37] The process of concentration of ownership and production within the defense industry has continued from the national to the international level. As a result, a limited number of extraordinarily large companies—Boeing, General Dynamics, Lockheed Martin, Northrop Grumman, and Raytheon in the United States; and BAE Systems, EADS and Thales in Western Europe—have emerged, each producing military goods and services for an annual value ranging from $5 billion to $19 billion.[38]

The SIPRI study on the trend of concentration in the defense industry shows that since the mid-1990s, the large arms-producing companies emerged to grow in size and improve their capability to acquire arms procurement contracts, through take-overs, mergers, joint ventures and other forms of company-to-company cooperation, both nationally and internationally.[39] The study notes that these devel-

---

[36] (*SIPRI Yearbook* 2002: 327)

[37] (ibid.)

[38] (*SIPRI Yearbook* 2002: 353)

[39] (op.cit., p. 352)

opments of concentration combined with the process of commercialization and privatization are resulting in fundamental changes in the global system of arms production and arms trade.

Table 4. Change in Concentration Ratios, SIPRI Top 100 Companies, 1990-2000 (%)

Company Sections	Concentration Ratios (% of combined total of Top 100)					
	Arms Sales			Total Sales		
	1990	1995	2000	1990	1995	2000
5 Largest Companies	22	28	42	33	34	40
10 Largest Companies	37	42	58	51	53	57
15 Largest Companies	48	53	66	61	65	68
20 Largest Companies	57	61	72	69	73	76

Source: The SIPRI Arms Industry Database (SIPRI Yearbook 2002, p. 327)

U.S. Predominance and 'Unilateralism' in International Defense R&D Cooperation

Throughout the post-Cold War nationalization and consolidation of the defense industry, the U.S. predominance in defense R&D and production has increased dramatically. The U.S. government has an outstanding level of total defense expenditures and particularly defense R&D expenditure (see **Tables 5** and **6**).

Table 5. Military Expenditure 2001 (in U.S.$ billion)

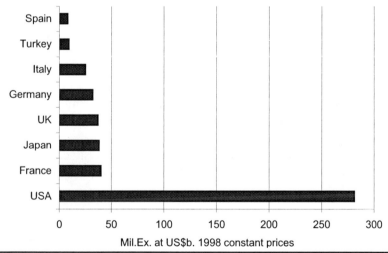

Source: SIPRI Yearbook 2002

Table 6. Government Expenditure on Military R&D 1998
(billion U.S.$)

Italy	~100
Spain	~300
Japan	~500
Germany	~1000
France	~3000
UK	~3500
USA	~40000

in US$ at constant 1998 prices

Source: SIPRI Yearbook 2002

Not surprisingly, U.S. predominance in the defense industry also is profound. According to SIPRI's "100 largest arms-producing companies, 2000,"[40] U.S. companies account for 43 out of the 100 largest arms-producing companies in the world (see Figure 6). In the same year, U.S. companies shared 60% of total sales for the top 100 arms-producing companies (see **Figure 7**).

Due to the predominance in quality and quantity of defense R&D and production, the U.S. defense industry is much less susceptible to challenge from foreign defense companies. As a result, the U.S. defense industry can develop and implement its own strategy and interest calculus with respect to international defense business. First, the U.S. defense industry conducts its business in an increasingly pragmatic manner for cost-effectiveness. One effective way to increase cost-effectiveness is to lessen bureaucratic impediments to arranging international defence business.. Thus, major U.S. companies increasingly emphasize company-to-company coordination, bypassing complex government-to-government talks and coordination. Second, as

---

[40] (*SIPRI Yearbook* 2002: 357-363)

more priority is given to cost-effectiveness in the post-Cold War defense market, international defense business is increasingly free from the traditional framework of alliances that was mostly established during the Cold War. For instance, in the JSF combat aircraft program, the United States accepted participation of Singapore as a non-traditional partner.[41] Third, reflecting the U.S. predominance in the international defense business, the United States often leads an international defense R&D project in a pragmatic way to secure its own interest. For instance, in case of the JSF program, the United States ranked its foreign partners' participation (Level 1 ~ Level 4) in terms of the degree of financial and technological contribution that a foreign partner provided. Fourth, in organizing international defense R&D programs, the United States, knowing its technological predominance, requests possible foreign partners to adopt U.S. standards of arms export controls. A typical example of this arrangement is the Defense Trade Security Initiative (DTSI) that requires foreign partners to adopt U.S. standards of arms exports control in return for easier transfer of U.S. defense technology to the foreign partner in question.

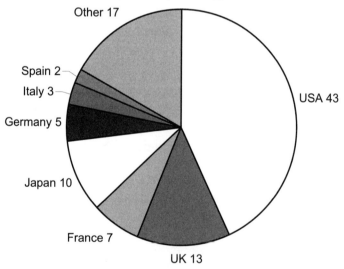

Figure 7. The 100 Largest Arms-Producing Companies, 2000
(number of companies by country)

Source: Calculation based on 'The 100 largest arms-producing companies in the OECD and developing countries, 2000', SIPRI Yearbook 2002: 357-363.

[41] 'U.S. offers new JSF participation to Singapore', *Defense News*, 19-25 August 2002, p. 3.

Figure 8. Share of Total Arms Sale for Top 100 Arms-producing Companies, 2000

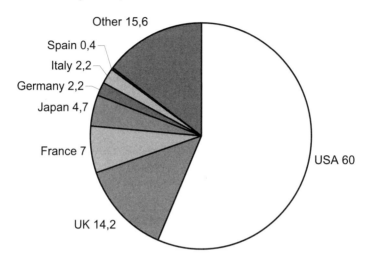

Source: Calculation based on 'The 100 largest arms-producing companies in the OECD and developing countries, 2000', SIPRI Yearbook 2002: 357-363.

International Defense R&D Cooperation: From Competition to True Cooperation

Trans-Atlantic and Trans-Pacific Convergence in Defense R&D Cooperation

The harsh realities of the post-Cold War defense market have strengthened market principles even in defense business; far more emphasis and priority is given to maximizing profits rather than the geo-political benefits of international armaments cooperation. The selection of foreign defense partners for international programs also departs from the traditional framework of Cold-War alliances. The predominance of the U.S. defense industry enhances this trend and as a consequence, a kind of American 'unilateralism' has emerged in international defense industrial affairs. During the Cold War, government-to-government coordination and bureaucracy consumed much time and energy in the case of Transatlantic armament cooperation within the NATO framework. Currently, it is the United States that usually takes the decisive lead in arrangement of international defense R&D cooperation; since the United States is virtually the single decision-maker, there is no need for too much government-to-government bureaucracy, which is time-saving and makes a cooperative program more cost-effective.

A typical example would be the JSF program, the JSF is the first U.S. combat aircraft project that attempts to meet the needs of three U.S. military services and foreign customers for a common platform.[42] However, although it is international and led by the United States,"[f]oreign industrial participation has been accepted only according to 'best value' based on competition. In contrast to most earlier international programs, there is no *juste retour* principle, (i.e. guaranteed industrial involvement in relation to the national financial contribution) and no offset arrangement."[43] Under these conditions, foreign countries are given opportunity to be involved in the program at different levels of involvement and costs (as indicated by levels 1–3 plus a fourth 'major participant'). Full partner participation (which requires paying 10% of the costs) permits direct influence on requirements; to date, only the UK is a full partner. Associate partners (2–5% of the cost) can influence the requirements for the conventional take-off and landing variant; Informal partners (1–2% of the cost) cannot influence the requirements; and the 'security cooperation participant' (via the Foreign Military Sales program) will receive extensive unclassified and non-proprietary information needed to evaluate JSF as a possible future acquisition.[44] The JSF international program is managed on an industry-to-industry rather than a government-to-government basis; teaming and subcontracting industrial relationship will be solely developed between U.S. weapon system contractors and country industrial partners (*NATO's Nations and Partners for Peace*, April 2002, p. 166-167).

The JSF program will likely become a new model, not only for Trans-Atlantic armaments cooperation but also for other international defense R&D program arrangements as well. For instance, international R&D cooperation for missile defense systems appear to be organized in a similar way as the JSF program, albeit with many indi-

---

[42] (*SIPRI Yearbook* 2002: 397). Designs for the U.S. Air Force, Navy and Marine Corps as well as the British Royal Air Force and Navy are to be produced from the same production line, with initial market of over 3000 aircraft, replacing the A-6, A-10, AV-8B, F-16 and to some extent F/A-18 combat aircraft in the U.S. and the harrier and Tornado in the UK.
[43] (*SIPRI Yearbook* 2002: 396).
[44] (*SIPRI Yearbook* 2002, p. 397).

vidual different levels and types of participation for each foreign partner.[45] Currently, Japan is involved in the development of the SM-3 missile for AEGIS-based (BMD). But European allies also are joining the development of a variety of missile defense systems, either entire missile defense systems, sub-systems or components, or merely to receive routine briefings.[46] The interests in missile defense among the European NATO states are increasing and the United States is willing to attract European partners for future co-development in a variety of missile defense systems.[47] Given the different kinds of cooperation involved, the United States may sign bilateral agreements with individual European countries for each missile defense program. It is not just missile defense programs that are following the new model; even development of next generation maritime patrol aircraft may be arranged in a similar way to the JSF program.

Future Prospects: Toward True and Better Cooperation

In sum, new models of organizing international defense R&D cooperation are moving beyond the framework of the Cold War alliance, are dynamic in nature and may potentially converge. In the new post-Cold War framework, international defense R&D programs shift the emphasis to cost-effectiveness, timesaving, and maximizing profit in order to survive in an increasingly saturated global defense market. More privatization and consolidation of the defense industry is inevitable. This implies that giant private defense companies will increasingly steer global defense businesses as primary actors while governments will assume roles as coordinators to approve licensing, supervise technology transfer issues, and the like.

---

[45] 'Seeking partners to develop and deploy a global missile defense system, the United States is offering European countries a variety of roles, ranging from industrial cooperation to financial contributions, a senior U.S. administration official said. "It is similar but not identical to the JSF program . . . where allies can design their levels of participation . . . the degree to which what they want to be involved is up to them" ' (*Defense News*, 19-25 August 2002, p. 1).

[46] (*Defense News*, 19-25 August 2002, p. 1).

[47] 'A U.S. delegation led by Assistant Secretary of Defense for international security policy J.D. crouch briefed the North Atlantic Council of ambassadors of NATO's 19 nations on July 18 on the possibility of European industrial participation in missile defense. This was the first time Europeans were offered the possibility of such cooperation' (*Armed Forces Journal*, November 2002, p. 36). Also, 'Allies Rethink BMD: New Threats modify European Resistance' in *Defense News*, Aug 26-Sep 1, 2002, p. 15.

Another implication is the 'Japanization' of defense R&D strategy; i.e., the application of civilian dual-use technology to military systems, due to diminishing government subsidies and constrained defense expenditures. While active utilization of civilian dual use technology is an efficient way to save on costs, this trend also will result in a higher risk of proliferation of sensitive technologies. Given these two new dynamics, the role of government should be strengthened as a guarantor of transparency and accountability in defense procurement policy-making and as a primary actor to better control proliferation of sensitive technologies, both military and dual-use.

Emerging U.S. 'unilateralism' in global defense businesses is another issue that requires careful scrutiny and better handling. The U.S.-Japanese partnership in defense R&D cooperation has always had one clear feature—complementary partnership. Japan, knowing its limited defense technological capability and resources, has satisfied itself to be a complementary partner of the United States, although Japan maintained *kokusanka* ('indigenization') efforts in some distinct areas such as electronics and missiles. Thus, for the Japanese defense industry, U.S. 'unilateralism' has long been a fact with which to deal. On the other hand, for the European defense industry, U.S. unilateralism is a fact that they understand as being inevitable but difficult to accept. While European companies are eager to get involved in U.S. defense R&D programs for cutting-edge technologies, the European defense industry has tried to maintain 'European' autonomy for less-innovative weapons systems by enhancing regional cooperation. Apparently the European defense industry and governments are frustrated by recent U.S. 'unilateralism' and have incentives to resist this trend. If the United States takes too much advantage of its predominance, a dangerous political 'backfire' may result which could harm the security of both the United States and Europe in the longer term.

As an example, the EU has decided to welcome China's investment of more than €230 million in its Galileo project—the EU's rival project to the Pentagon-controlled Global Positioning System.[48] Experts point out that this will enhance defense cooperation between Beijing and Brussels by opening China's growing market to the European

---

[48] The Galileo is a €3.25 billion project will compete with the U.S. in navigation satellite technology (*Financial Times*, 19 September 2003).

defense industry and by providing China with a more sophisticated satellite system, which would eventually help improve China's guided missile technology with military hardware application. China has not joined (and is unlikely to join) the Missile Technology Control Regime (MTCR) or subscribed to the International Code of Conduct (ICOC) against ballistic missiles proliferation. This means that increasingly sophisticated ballistic missiles will likely proliferate from China—a major exporter of ballistic missiles—to states of concern or involved in conflict, which would eventually threaten European security in a long term. This is particularly ironic, given the fact that EU was most helpful in multi-nationalizing the MTCR into the ICOC. Partly to blame is the lack of consistency in the EU's actions and various policy implementations. Arguably, if the United States was more open to share the benefits of its predominance in the GPS system with its European allies, such a strategic backfire could have been prevented. According to an ancient proverb, when "two parties are fighting, it is always the third party who wins the game." Both the U.S. and European defense industries may do well to remember this proverb in planning international defense R&D programs in the future.

Appendix A

Symposium Brochure
And
Agenda

Appendix A

Symposium

European Defense Research & Development
- New Visions & Prospects for Cooperative Engagement -

June 6, 2003

Symposium Chairman – Jeffrey P. Bialos

Center for Transatlantic Relations
The Paul H. Nitze School of Advanced International Studies
1717 Massachusetts Avenue, NW
Suite 525
Washington, D.C. 20036-1984
Telephone 202.663.5880
www.transatlantic.sais-jhu.edu

Symposium

European Defense Research & Development
- New Visions & Prospects for Cooperative Engagement -

Venue:
Center for Transatlantic Relations
1717 Massachusetts Ave., NW
Suite 525
Washington D.C. 20036
202.663.5880

June 6, 2003

7:45 a.m.

Continental Breakfast

8:00 a.m. – 8:30 a.m.

Chairman's Welcome & Introductory Remarks: Framing the Issues

- Jeffrey P. Bialos, Executive Director, Program on Transatlantic Security & Industry, Center for Transatlantic Relations, Johns Hopkins School of Advanced International Studies, Washington, D.C.

8:30 a.m. – 10:15 a.m.

Panel 1: European "National" Directions on R & D – New Processes & Approaches

This panel will address the new directions and approaches that key European governments are taking on defense research and development. What are the evolving European "national" concepts of transformation and net centric warfare, and how are European government's research and development seeking to translate these new approaches into practical solutions for the war fighter?

- David Gould, Deputy Chief Executive, Defense Procurement Agency, United Kingdom Ministry of Defense
- Major General Staffan Nasstrom, Chief Operational Manager, Defense Materiel Administration, Swedish Ministry of Defense
- Yves Boyer, Assistant Director, Fondation pour la Recherche Stratégique, Paris, France & Chairman, French Society for Military Study

- *Panel Chair: Dr. Jacques Gansler, Professor and Roger C. Lipitz Chair, Center for Public Policy & Private Enterprise, University of Maryland, College Park, Md.*

10:30 a.m. – 12:30 p.m.

Panel 2: Leveraging European Dual-Use Technologies for Defense Needs

This panel will address Europe's leveraging of commercial "dual use" research in aerospace, space and related areas for defense. It will consider new and important areas of focus, approaches, and visions, including the prospect of new European Union responsibility and programs.

- Sir John Chisholm, Chief Executive Officer, QinetiQ Group Plc., Farnborough, the United Kingdom
- Klaus Becher, Associate Research Fellow, EU Institute for Security Studies, Paris France
- Dominique Vernay, Corporate Technical Director, The Thales Group, Paris France
- Daniel Hernandez, Director of Research, Centre National d'Etudes Spatiales, Paris, France

- *Panel Chair: Professor Kenneth Flamm, Director, Technology and Public Policy Program, Lyndon B. Johnson School of Public Affairs, University of Texas at Austin*

12:45 p.m. – 2:00 p.m.

Lunch

Luncheon Address:

Dr. Anthony Tether
Director, Defense Advanced Research Project Agency
The Pentagon, Washington, D.C.

2:00 p.m. – 3:30 p.m.

Panel 3: European R & D Cooperation

This panel will address existing modes and subjects of European cooperation in the armaments research and development field (including the Letter of Intent agreement, Europa Memorandum, and the European Technology Acquisition Program), and prospects for enhanced coalescence through the European Union (including the prospect of an EU Armaments Agency). Are these modalities and approaches effective, and will they facilitate European transformation and capability enhancement in support of NATO and EU Headline goals?

- Dr. Ernst Van Hoek, Chairman, Western European Armaments Group, The Hague, the Netherlands
- Daniel Deviller, Senior Vice President, Industrial Research & Technology & Chief Technology Officer, EADS, Paris, France
- Andrew James, Senior Research Fellow, Policy Research In Engineering, Science & Technology, University of Manchester, Manchester, United Kingdom

- *Panel Chair: Professor Gordon Adams, Director of Security Policy Studies Program, George Washington University, Washington D.C.*

4:00 p.m. – 5:30 p.m.

Panel 4: Trans-Atlantic R & D Cooperation: Top-Down & Bottom Up Realities & Possibilities

The panel will explore the prospects of impediments to Transatlantic collaboration, including lessons learned. It will identify potentially productive areas of and modalities for future collaboration with a view toward achieving transformation, standing up the NATO Rapid Reaction Force and otherwise promoting inter-operability and enhanced capabilities among allies. The "top down" refers to government cooperative programs and "bottom up" to industrial-driven activities.

- Kenneth Peebles, Director, NATO Research & Technology Agency, Paris, France
- Spiros Pallas, Consultant, Washington, D.C.
- Dr. Masako Ikegami, Associate Professor & Director, Center for Pacific-Asia Studies, Stockholm University, Sweden

- *Panel Chair: Adrian Kendry, Senior Defense Economist, NATO, Brussels, Belgium.*

5:30 p.m. – 6:00 p.m.

Concluding Remarks

- Jeffrey P. Bialos, Symposium Chairman

Appendix B
Sweden's Approach to Defense Research and Transformation

Major General Staffan Näsström

Appendix B

Sweden's Approach to Defense Research and Development & Transformation

Presentation at the Center for Transatlantic Relations at
John Hopkins University
"European Defense Research and Development:
New Visions and Directions"

Washington, D.C
June 6, 2003

Agenda

- Background
- The Swedish Defense transformation program and status June, 2003
- Sweden's focus on co-operation
- Sweden's focus on R&T/D
- Sweden's approach for an efficient and secure procurement process
- Conclusions

Appendix B 149

Background

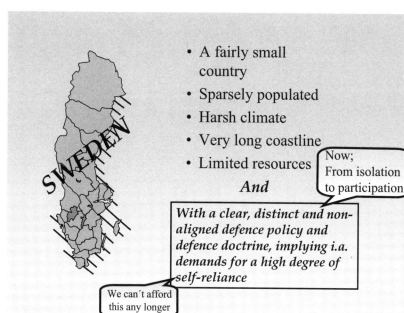

- A fairly small country
- Sparsely populated
- Harsh climate
- Very long coastline
- Limited resources

Now; From isolation to participation

And

With a clear, distinct and non-aligned defence policy and defence doctrine, implying i.a. demands for a high degree of self-reliance

We can't afford this any longer

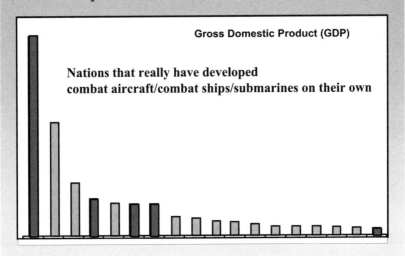

What are the characteristics of a successful defense project?

- 1. Existence of a clear defense doctrine/purpose/role
- 2. Real demand of products is identified
- 3. Governments and authorities capable of long-term and persistent decisions
- 4. Long-term R&T programs have been launched in support of the projects
- 5. Creation of a constructive attitude to the activities attracts the front rank recourses

Appendix B 151

What are the characteristics of a successful defense project (cont.)

- 6. Stakes by strong competent industrial groups
- 7. Creation of interconnection between military and civilian production
- 8. Exploitation of competition
- 9. A very competent and qualified customer

The Swedish defense transformation program and status June 2003

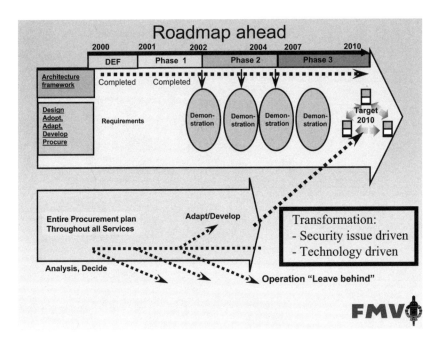

Legacy systems defined as services

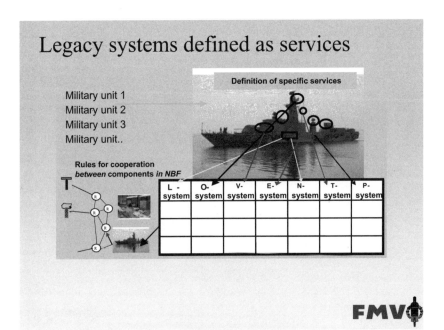

Appendix B 153

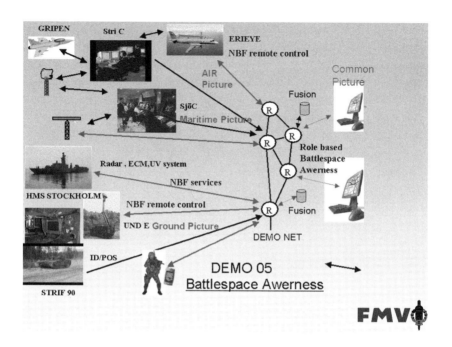

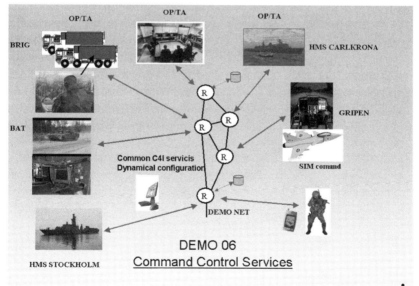

Sweden's focus on cooperation

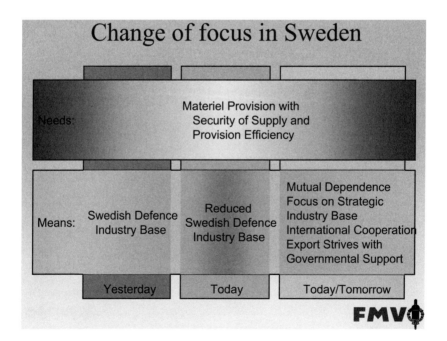

Sweden's cooperation strategies

Security of supply	Transfer and export procedures	Security of classified information	Defense related R/T	Treatment Of Techn.it	Harmonization of Mil.req.

- *Strategies*
  - *Focus on R&T/D, strategic competence and acquisition competences. This is the bottom line*
  - *Focus on realization and full implementation of Framework Agreement*
  - *Secure the transatlantic link and other important links*

US-EU cooperation situation

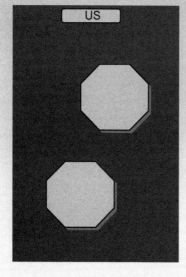
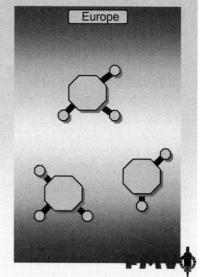

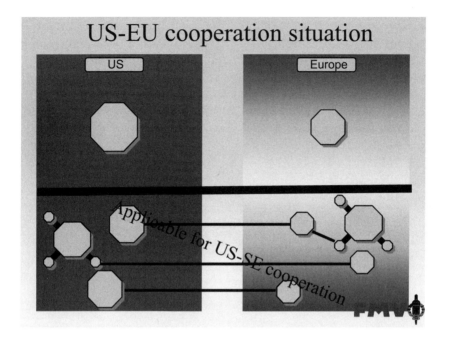

Sweden's focus on R&T/D

Appendix B   157

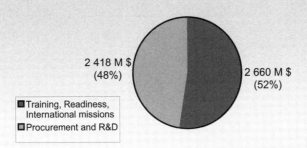

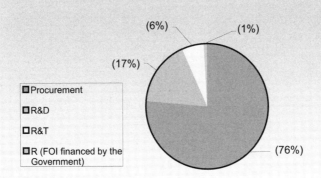

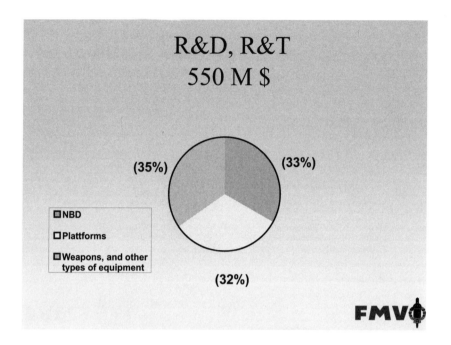
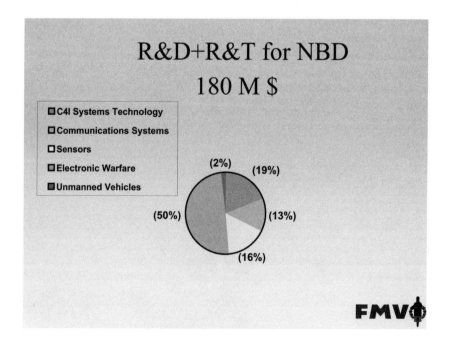

Sweden's approach for an efficient and secure procurement process

General key issues for all

- To find the genuine needs of products in a changing and turbulent world
- To dare start up new projects
- To find the right timing for cooperation on all levels and for all relevant bodies

Five crucial prerequisites

- Harmonize military requirements
- Harmonize processes
- Use M&S as a facilitating and cooperation tool
- Focus on R&T/D
- Cooperate

Industry strategy

- Defence industries with core competences *inside* Sweden, crucial to the Armed Forces
  - Education
  - Adjustments, development and smart use etc
- Ownership important to the Armed Forces
  - Strategically and strong
  - Genuinely competent and interested
  - *Not* necessarily Swedish
- Leadership important to the Armed Forces
  - Profitable
  - Competitive
  - Surviving
  - Supporting
- Therefore important with good opportunities and comprehensive infrastructure *inside* Sweden for industry

Conclusions

A new security situation

- A new security situation enabled us to review our strategies regarding industrial base
  - Accept consolidation
  - Accept cooperation
  - Accept foreign ownership
- All built on interdependence and security of supply from a Swedish perspective

What is crucial for Sweden

- To focus and take long-term decisions as before
  - Now with governmental partners as well as industrial partners
  - On products that also other Armed Forces need
- To foster and contribute to open and transparent processes inside Europe and between US and Europe
- To find areas where Sweden will become world champion – NCW !?

Thank you for listening

Staffan Näsström

Defense Materiel Administration Sweden

Appendix C
Dual Use Technology in European Space Research and Development

Daniel Hernandez

Appendix C

Dual-Use Technology in European Space Research and Development

Daniel Hernandez

The application of satellites can be now found in many domains from fundamental physics to commercial utilization and we currently depend every day on space for many of our activities: transportation, communication, culture and entertainment, and security.

Both militaries and civilians thus are depending more and more on space systems. Consequently, more countries are considering space as strategic and therefore desire developing an indigenous capability so that they would not depend on other countries good will to get the space systems they wish to set up. But the development of a space capability is very expensive; some countries have the hope that they could access the international commercial market to lighten the cost of maintaining this space capability.

Inevitably, with the growing number of suppliers comes growing competition. The example of the launcher situation is illustrative:

- In 1980, only one country (U.S.) was offering to launch foreign satellites but with constraints;
- In 1990, only two suppliers where possible (U.S. and Europe, as USSR was not present on the commercial scene);
- In 2000, a number of solutions were possible: U.S., Europe, India, Russia, Ukraine, Japan, China, and Israel were able to launch satellites and had or were ready to have a commercial activity on the international scene; and
- Today, several others countries are or could be developing launching capabilities (Brazil, North Korea, Pakistan and possibly others).

In the technology of satellites, the situation is not very different especially with the possibility to use small satellites (micro satellites in general, in the range of 50 to 100 kg) for a reduced amount of resources.

Several factors together support this process of multiplication of space system sources:

- the threat of dependence on other countries' decisions to acquire a space system pushes those believing that space is a strategic asset to develop some form of independence;

- The need, for those having a space production capability, to export if they are not ready to carry alone the burden of funding this expensive investment forever; and

- The compensation requested by some countries at an allocation of a contract, under form of an assistance to develop a space activity (for instance small satellites).

Note: this is not specific to space; it is also the case for many governmental contracts in the armament and commercial aerospace sectors.

The process seems to accelerate with more and more sources on the market, i.e. not only more countries but also in the United States, Russia and Europe more suppliers (recent examples in Europe: launcher VEGA and Radar Sar-lupe).

Most suppliers have access to some markets under privileged conditions: they may build on these markets their know-how and tools with protection and help from their governments. Access to open markets gives them the possibility to better amortize their investments; they even can apply marginal costs.

The institutional contracts are usually reserved to national companies. For these contracts it is usually distinguished to type of governmental customers: militaries and civilians. The following table gives an idea of the size of these institutional programs in the United States and Europe.

	Civilian space	Military space	Institutional space Total institutional space	Ratio civilian/ military	Gross domestic product (GDP)	Ratio Institutional space activity/GDP
USA	16.2	15.6	32	1	11,000	3‰
Europe	5.3	0.6	6	9	9,000	0.7‰
Ratio USA/ Europe	3	26	5.3		1.6	

Note: Figures in B$ or B€, or no unit (when ratios)

It is well known that the U.S. government is doing much more in space than European ones: 3 times more in the civilian domain and 26 times more in the military domain. Thus, it is not surprising that U.S. industry is much more powerful than its European counterpart. For Europe, the export market is essential; it is less economically important for US suppliers.

In the United States, the civilian and military domains represent equivalent sizes. However, this is not the case in Europe where the civilian space activity is nine times more important than the military one. Thus, there is no need to say that the influence of military contracts is therefore not the same in the United StatesUSA and in Europe. In Europe the military domain is benefiting more from the civilian space activity than in the United States. Nevertheless it seems clear that, in space, both military and civilian domains benefit from the each other.

The first important synergy is related to the fact that if each domain had to pay the full cost of acquiring and maintaining the know-how and the equipments and installations necessary for the study, development and operations of their own space programs, they would do far less. Every space program is benefiting from any other, no matter if they are similar in term of missions and technologies. As an example one could mention the scientific space programs that are feasible at the current conditions only because of the commercial programs.

More directly, most space technologies have dual uses, such as:

- *Launchers.* All launchers are able to launch satellites without consideration to their mission. But it is fair to say than some orbits are not used by military programs such as those for planetary exploration.

- *Components, equipments.* This includes the satellites busses and some parts of the payloads as well as components (e.g., batteries, sun sensors, detectors).

Military Utilization of Space

Let us consider the various military satellites systems and more in detail the telecommunications systems, and identify the difference between civilian and military programs.

Telecommunications

Several countries in Europe have military telecommunication satellites, including the UK (Skynet), France (Syracuse), Spain (Hispasat) and Italy (Sicral). There is *in principle* no major difference between civilian and military basic needs and consequently the technologies. *However,* besides the basic specification, military systems have generally several differences:

- Frequencies specific to military uses (for instance the 7-8 GHz bands);
- Capacity to cover large zones and to offer spot beams;
- Protection against jamming whereas civilian are only protected against interferences (jamming is an intentional interference);
- Protection against unauthorized access to telecom service, to telemetry and telecommand of the satellite; of course, some degree of protection is also required by commercial systems that need to safeguard their investment and protect the transactions of their users; and
- Protection against physical aggression (radiation).

For the most part, these capabilities are not requested in civilian programs but the basics technologies used are the same.

Eavesdropping

This application is specific to militaries

Alert

This application is specific to militaries

Space Watch

The need to observe the sky to identify and monitor the environment of the orbit is not specific to military domain. The civilian programs need to know any risk for instance of collision with other satellites or debris. The military organizations require in addition the analysis of the various objects in orbit to characterize them especially if they could be unfriendly.

Navigation

The need is basically the same in both types of applications but militaries require higher degree of protection against interference or jamming and require also the possibility of denying use of the system to non friends. The accuracy of localization could also be better for military uses.

Earth surface observation

Imagery of the earth is needed by both types of users but militaries are requesting higher accuracy, shorter response time to a demand, all weather and time access plus higher encryption. The civilian would like to get similar performances but the pressure for getting satisfaction to their need (or capacity to afford) is lower, therefore they have systems performing like the ones the militaries had 10, 15 or even 20 years ago.

Meteorology

No big differences between both domains.

Others

There are several other applications of space for defense users, including oceanography, altimetry, and geology. However, there is no special difference between civilian and military demand on the space segments characteristics.

Civilian Utilization of Space

In addition to the mission seen in the preceding paragraph, the civilian domain includes several other type of utilization of space not currently of interest to militaries. These include, for example, the manned missions, microgravity experiments, exploration of the solar system, and astronomy. These missions require several technologies rather a single specific one, including energy production and storage far from the sun, and orbit control.

But some specific needs of these missions are relying on technologies developed for military missions. The best example is certainly the infra red detector technology needed for vulcanology or astronomy. Science would not be able to afford the development of specific detectors if it were not the militaries that paid for the major investments; defense ministries need this technology for night vision instruments.

Future Space Programs

Let us guess what could be the future of space in Europe. France is by far the first European investor in space. The French space agency—CNES—is reporting to both ministries "Defense" and "Research and New Technologies". Both are funding the national space activity: Defense for 0.45 B€, Research for 1.3 B€. Indeed, *from the very beginning, civilians and militaries have been working together in space: on launchers and missiles, satellites (Spot-Helios, Telecom 1 and 2, etc.) and technologies.*

Other European countries have lower budgets for space and if most of them have a civilian activity very few have a military space program.

Today, and maybe the recent war in Iraq has evidenced that, there is no real European foreign policy and common defense, but the European Union is taking steps in this direction every day. The people in Europe are feeling increasingly bound together if not yet fully part of a same "boat". There is no doubt that tomorrow not only civilian projects will be conducted with participants from various European countries but military programs—space and non space—as well.

CNES, which reports to the Ministry of Defense as well as the Research ministry, is participating to the development of military programs done in France under DGA (*Délégation Générale Pour l'Armement*) (already referenced in prior chapters) responsibility. As a consequence, when establishing its R&D plans, CNES is coordinating with Defense minister representatives, at various steps of the process. Let us recall that the process for establishing the CNES R&D plan is the following:

1. long term vision for missions
2. long term vision for technologies
3. identification of keys technologies
4. orientation for the calls for ideas
5. R&D plans updated yearly and entirely rebuilt every three years

Generally, with several exceptions, a number of key elements—"faster, better, cheaper" are driving the evolution of the space missions. Specifically, these include:

- Faster mission development;
- Better performance such as better resolution, wider coverage, shorter access time;
- Lower mission costs;
- Closer to the user needs; and
- Better integration of space systems in networks.

At the level of technology, the possible general trends are:

1. Use of COTS for electronic parts as well as software and hardware;
2. Decrease the constraints induced by launch to ease the design and manufacture of satellites;
3. Improve knowledge in space environment (e.g., radiation, debris, atomic oxygen) so as to adjust the design margin to the minimum possible and conceive measures to protect the satellites to sudden risks;
4. Improve propulsion efficiency (fuel represents a large share of the satellite weight);
5. Improve energy production and storage (energy is also repre-

senting a large weight and in addition represents a large cost of the satellites;

6. Improve reliability and life time;

7. Improve autonomy and survivability of space systems;

8. Improve protection of environment (for instance production of debris in orbit) and entering into more precise items;

9. More and more digital systems with more and more computers and memories combined with progresses in electronic integration leading to continuous decrease in size, weight, power consumption;

10. detectors more sensitive; and

11. Electronic able to operate at higher frequencies.

All these points are common to any type of mission, civilian or military. As noted above, most technologies used in space systems are not different when analyzed at a sufficiently granular level. Therefore, synergy between military and civilian projects is high. When analyzed at the system level, some military and civilian missions show differences. For instance the Herschel (European astronomy satellite) is using a telescope with 3.5 m diameter mirror and the American James Webb Space Telescope will use a 6.5 m diameter telescope. No doubt that these technologies could well lead to earth observation satellites with high resolution. For instance, a geostationary satellite would be able to image the Earth with resolution in the meter range. This would mean a capacity to very quickly (in terms of minutes) observe a given zone on the Earthwith many details. Formation flying, i.e. synthesis of large instruments through a number of small satellites flying together, is an elegant solution for astronomy (very large telescope for very high resolution) but also for earth observation (radar or optical) or accurate localization of transmitters.

To some extend, military systems of today will lead to-morrow to civilian systems but less frequently to the contrary. This is due essentially to the fact that governments are, in general, less ready to accept paying for expensive projects dedicated to science than to defense. If the pressure for having a satellite system better than what is already available on the market is great, then it is necessary to pay for development of new technologies. And this is more frequent for military systems than for civilian ones (at least in space, but some counter-example exist). In that respect the situation may be slightly different in the United States and in Europe when considering the amount of expenses for space systems by civilians and militaries.

Economic Impact of R&D

According to many, if not all, economists that follow the way opened by Joseph Aloïs Schumpeter in the first half of past century, R&D plays an essential role, through improvement on competitiveness, on economic growth and development. The Organization for Economic Cooperation and Development (OECD) is one of the organizations monitoring importance of R&D in the world and its effect on economy. Hereafter abstracts from two OECD documents illustrate this activity. For instance it establishes that from 1980 to 1998 in 16 major countries:

- "... an increase in R&D of 1% in business R&D generates 0.13% in productivity growth ...";
- "... 1% more in public R&D generates 0.17% in productivity growth ..."; and
- "... the negative effect of the share of defense in public R&D budgets, as it is not the main purpose of defense R&D to increase productivity ..."

OECD and other studies suggest that governmental funded R&D is very important for economy of a country and on average the benefits offset largely the cost of this R&D. But the more general the R&D is, the more beneficial it is for economic growth. And of course the more general the R&D the more dual it is.

Conclusion

Without the civilian activity, the space military programs would not be what they are today and vice versa. It is the case for many different domains and not only the case for space. Generally speaking, technology itself is not specific to the defense or civilian sector. Rather, it is the use of the technology that gives it a military or civilian orientation.

About the Authors

Klaus Becher—Mr. Becher is currently the first Associate Research Fellow at the European Union's Institute for Security Studies (ISS) in Paris, covering space policy issues, and also the founder and managing partner of the consultancy firm Knowledge and Analysis LLP located in Surrey, England. From 1999 to May 2003, he was the Helmut Schmidt Senior Fellow for European Security at the London-based International Institute for Security Studies where he co-directed the IISS/CEPS European Security Forum in Brussels. Previous to this, Mr. Becher spent two years as deputy head of the Strategic Affairs, Arms Control and Technology Department at the *Stiftung Wissenschaft und Politik* (SWP) in Ebenhausen, Germany. From 1988 to 1997, he held a sequence of positions at the German Council on Foreign Relations (DGAP), then based in Bonn, including secretary of the International Security Studies Group and executive editor of the yearbook *Die Internationale Politik*. Mr. Becher is married, has three children and lives in Bonn and London.

Jeffrey Bialos—Mr. Bialos is currently the Executive Director of the Transatlantic Security and Industry Program at the Johns Hopkins University's Paul H. Nitze School of Advanced International Studies Center for Transatlantic Relations and a partner in the law firm of Sutherland, Asbill & Brennan. Mr. Bialos previously served in a number of senior positions in the Clinton Administration, including most recently as Deputy Under Secretary of Defense for Industrial Affairs, in which position he received the Department of Defense's Distinguished Service Medal. He has published numerous articles and prepared numerous reports on defense and security issues, and conducted studies of a variety of issues concerning the transatlantic armaments market, NATO, and related subjects. He also was appointed by Governor Mark Warner of Virginia to serve on Secure Virginia, a panel overseeing Virginia's homeland security efforts. Mr. Bialos is a graduate of Cornell University (A.B. with honors), the University of Chicago Law School (J.D.), and the Kennedy School of Government, Harvard University (M.P.P).

Dr. Yves Boyer—Dr. Boyer is currently deputy director at the *Fondation pour la Réserche Stratégique*, a think-tank in Paris. In his various capacities, he chairs a working group of the French Ministry of Defense's Defense Science Board dealing with military R&D as well as the *Société Française d'Études Militaires*. Prior to this he was a senior researcher at the French Institute for International Affairs (IFRI), a former researcher at the International Institute for Strategic Studies and a Woodrow Wilson Scholar. He is a member of the editorial boards of *Annuaire Français* de *Relations Internationales*, the *International Spectator* (from the IAI in Rome), the *Revue de Géoéconomie* and the *Questions Internationales* (in Paris) and also is an assistant professor at the French army academy.

Dr. Kenneth Flamm—Dr. Flamm has been Professor and Dean Rusk Chair in International Affairs at the Lyndon B. Johnson School of Public Affairs, University of Texas at Austin, since the fall of 1998. Since 1995, he has also been at the Brookings Institution as a Senior Fellow in the Foreign Policy Studies program, a position he also held from 1987 to 1993. From 1993 to 1995, Dr. Flamm served as Acting Assistant Secretary of Defense (Economic Security), Principal Deputy Assistant Secretary of Defense and Special Assistant to the Under Secretary (Dual Use Technology Policy and International Programs), then as Special Assistant to the Deputy Secretary of Defense (Dual Use Technology Policy) and Principal Deputy Assistant Secretary of Defense (Economic Security). He was awarded the Department's Distinguished Public Service Medal by the Secretary of Defense. Dr. Flamm has taught at the *Instituto Tecnologico A. de Mexico* in Mexico City, the University of Massachusetts—Amherst, and at The George Washington University. He was awarded a BA with honors in economics from Stanford University in 1973 and a Ph.D. in economics from M.I.T. in 1979.

David Gould—Mr. Gould currently serves as Deputy Chief Executive of the UK Defense Procurement Agency. Previously, he has served in the UK Ministry of Defense since 1973, and his various duties have included NATO, nuclear plans and operations, and RAF equipment and logistics. He is a 1980 graduate of the NATO Defense College in Rome, and from 1983 to 1987 he was seconded to the Foreign and Commonwealth Office in the TAX Delegation to NATO. In 1992, he was appointed Under Secretary for Supply and Organization (Air Force), and in 1993 became Undersecretary for Policy in the MOD with responsibility for relations with former Warsaw Pact states and former neutrals. In May 1993, he was seconded to the Cabinet Office, Overseas and Defense Secretariat, with responsibility for contingency planning and current operations (including Bosnia, Counter-Terrorism, defense equipment, civil and military nuclear policy, export controls, and the implications of public use of cryptography. Upon returning to the MOD in 1995, Mr. Gould became Under Secretary (Fleet Support), with responsibility for Royal Dockyards and construction of the Vanguard Class and other nuclear submarines. Following the creation of a joint Defense Logistics Agency, he spearheaded the effort to establish the managerial structure of the new agency and the introduction of commercial-style accounting across Defense Logistics. In 1999, he became the Director General of Finance and Business Plans for Defense Logistics, and was appointed Deputy Chief Executive of the Defense Procurement Agency in July 2000.

Dr. Masako Ikegami—Dr. Ikegami is currently Associate Professor and Director of the Center for Pacific Studies (CPAS), Stockholm University. There she conducts interdisciplinary research on security policy and confidence building measures as well as arms control and disarmament in East Asia. She holds a Ph.D. in Peace and Conflict Research from Uppsala University and a Doctor of Sociology from the University of Tokyo.

Andrew James—Andrew James is a Research Fellow at the University of Manchester in the United Kingdom. His research focuses on the key industrial, technological and policy issues facing the aerospace and defense sector in four broad and complementary areas: the future of transatlantic defense industrial relationships and globalization; the consolidation of the aerospace and defense industry and the management of mergers and acquisitions; dual use and the commercialization of technology; and, defense science and technology policy and management. Mr. James has published more than 40 reports and papers on defense-related topics and has acted as a consultant to a range of corporate and government clients. Before joining the University of Manchester, Mr. worked as economic and industrial policy adviser to the Labour Party's Parliamentary Spokesperson on Industry in the House of Commons where he had a particular responsibility for analysis of the aerospace and defense sector as well as wider issues related to UK and European economic and industrial policy.

Stuart Koehl—Mr. Koehl is a defense analyst serving as a Fellow at the Center and focuses his research activities on Transatlantic security and industrial issues. A graduate of Georgetown University, he has conducted analyses for both government and industry on defense technology and transformation issues.